T0400768

ENVIRONMENTAL SCIENCE, ENGINEERING AND TECHNOLOGY

COAL COMBUSTION WASTE: MANAGEMENTAND BENEFICIAL USES

Environmental Science, Engineering and Technology

Additional books in this series can be found on Nova's website under the Series tab.

Additional E-books in this series can be found on Nova's website under the E-books tab.

Environmental Health - Physical, Chemical and Biological Factors

Additional books in this series can be found on Nova's website under the Series tab.

Additional E-books in this series can be found on Nova's website under the E-books tab.

ENVIRONMENTAL SCIENCE, ENGINEERING AND TECHNOLOGY

COAL COMBUSTION WASTE: MANAGEMENT AND BENEFICIAL USES

DANIEL D. LOWELL
EDITOR

Nova Science Publishers, Inc.
New York

Additional color graphics may be available in the e-book version of this book.

LIBRARY OF CONGRESS CATALOGING-IN-PUBLICATION DATA
Coal combustion waste : management and beneficial uses / editor, Daniel D.
Lowell.
p. cm.
Includes bibliographical references and index.
ISBN 978-1-61728-962-0 (hardcover)
1. Coal ash--Environmental aspects. 2. Coal ash--Recycling. I. Lowell,
Daniel D.
TD195.C58C6125 2010
621.31'21320286--dc22
2010027144

Published by Nova Science Publishers, Inc. ✣ New York

CONTENTS

PREFACE

Coal fired power plants account for almost 45% of electric power generated in the United States. The coal combustion process at those facilities generates a tremendous amount of waste. In 2008, industry estimates indicate that 136 million tons of coal combustion waste (CCW) was generated. That would make CCW the second largest waste stream in the United States, second to municipal solid waste, or common household garbage. How CCW is managed and how those management methods are regulated have come under increased scrutiny in recent years. This book examines the potential harm from CCW to human health and the environment, and delves into the investigation of the management of CCW which is essentially exempt from federal regulation.\

Chapter 1- In 2008, coal-fired power plants accounted for almost half of the United States' electric power, resulting in as much as 136 millions tons of coal combustion waste (CCW). On December 22, 2008, national attention was turned to issues regarding the waste when a breach in an impoundment pond at the Tennessee Valley Authority's (TVA's) Kingston, Tennessee, plant released 1.1 billion gallons of coal ash slurry. The estimated cleanup cost will likely reach $1.2 billion.

The characteristics of CCW vary, but it generally contains a range of heavy metals such as arsenic, beryllium, chromium, lead, and mercury. While the incident at Kingston drew national attention to the potential for a sudden catastrophic release of waste, the primary concern regarding the management of CCW usually relates to the potential for hazardous constituents to leach into surface or groundwater, and hence contaminate drinking water, surface water, or living organisms. The presence of hazardous constituents in the waste does not, by itself, mean that they will contaminate the surrounding air, ground, groundwater, or surface water. There are many complex physical and biogeochemical factors that influence the degree to which heavy metals can dissolve and migrate offsite—such as the mass of toxins in the waste and the degree to which water is able to flow through it. The Environmental Protection Agency (EPA) has determined that arsenic and lead and other carcinogens have leached into groundwater and exceeded safe limits when CCW is disposed of in unlined disposal units.

In addition to discussions regarding the potential harm to human health and the environment, the Kingston release brought attention to the fact that the management of CCW is essentially exempt from federal regulation. Instead, it is regulated in accordance with requirements established by individual states. State requirements generally apply to two broad categories of actions—the *disposal* of CCW (in landfills, surface impoundment, or mines) and its *beneficial use* (e.g., as a component in concrete, cement, or gypsum wallboard, or as structural or embankment fill).

Chapter 2- The U.S. Environmental Protection Agency's Office of Solid Waste (EPA OSW) is currently developing methods to evaluate the environmental, human health, and economic outcomes of specific EPA programs. As an initial step, OSW is examining the extent to which the costs and benefits of source reduction, reuse, and recycling may be quantified for a range of materials targeted by the Resource Conservation Challenge (RCC).

Coal combustion products (CCPs) are among the materials targeted by EPA's Resource Conservation Challenge (RCC). The RCC is designed to facilitate changes in the economics and practice of waste generation, handling, and disposal (e.g., by promoting market opportunities for beneficial use). Under the RCC, EPA has established three goals for increased beneficial use of CCPs:

Chapter 3- On December 22, 2008, the retaining wall broke on a waste retention pond at the Tennessee Valley Authority (TVA) Kingston Fossil Plant, Tenn., and an estimated 4.1 million m^3 of coal ash slurry was spilled onto the land surface and into the adjacent Emory and Clinch Rivers (TVA, 2009). This was the largest coal ash spill in US history. The coal ash sludge spilled into tributaries that flow to the Emory River and directly into the Emory River itself (Figure 1), which joins to the Clinch River and flows to the Tennessee River, a major drinking water source for downstream users. With funds provided by the Dean of the Nicholas School of the Environment of Duke University, in January 2009 our team began a preliminary investigation of the potential environmental and health effects of the spill. This preliminary work (Vengosh et al., 2009; Ruhl et al., in revision) has thus far revealed three major effects: (1) The surficial release of coal ash formed a sub-aerial deposit that contains high levels of toxic elements (arsenic concentration of 75 mg/kg; mercury concentration of 150 &g/kg; and radioactivity (radium-226 + radium-228) of 8 pCi/g). These pose a potential health risk to local communities as a possible source of airborne re-suspended fine particles (<10 μm). (2) Leaching of the coal ash sludge in the aquatic environments resulted in severe water contamination (e.g. high arsenic content) in areas of restricted water exchange such as the Cove area, in a tributary of the Emory River. Further downstream, in the Emory and Clinch rivers, much lower levels of metals were found due to river dilution, but with metals concentrations above the background upstream levels. (3) High concentrations of mercury in downstream sediments of the Emory and Clinch rivers indicate physical transport of coal ash in the rivers. The high concentration of mercury and sulfate in the downstream river sediments could impact the aquatic ecosystems by formation of methylmercury in anaerobic river sediments.

A recent survey of the amount of coal ash generation in the United States revealed that 500 power plants nationwide generate approximately 130 million tons of coal ash each year, 43 percent of which is recycled into other materials. The remaining 70 million tons are stored in 194 landfills and 161 ponds in 47 states (Lombardi, 2009). An EPA study (USEPA, 2007) identified 63 coal ash landfills and ponds in 23 states where the coal sludge is associated with contaminating groundwater and the local ecosystem. One of the major potential hazards of coal ash storage in ponds is the continuous leaching of contaminants and their transport to the hydrological system. As such, the TVA coal ash spill provides a unique opportunity to evaluate the large-scale impact of coal ash leaching on the environment and water resources.

Chapter 4- Chairwoman Johnson, Ranking Member Boozman, and members of the Committee. I appreciate this opportunity to discuss the coal ash spill at the Tennessee Valley

Authority's (TVA) Kingston Fossil Plant, the actions taken in response to the event, and our progress and plans for remediation of the site and protection of the environment.

The incident being discussed today occurred at TVA's Kingston Fossil Plant in Roane County, Tennessee. On behalf of TVA, we deeply regret the failure of the ash storage facility dike, the damage to adjacent private property in the Swan Pond community, and the impact on the environment. We are extremely grateful that no one was seriously injured.

TVA is committed to cleaning up the spill, protecting the public health and safety, and restoring the area. In the process, we will look for opportunities, in concert with the leaders and people of Roane County, to make the area better than it was before the spill occurred. This commitment will stand because TVA is part of the Kingston community through our employees who live and work there, and through the partnership of our historic mission to work for the economic progress of the Tennessee Valley region.

We are also committed to sharing information and lessons-learned from this event and the recovery with those in regulatory and oversight roles, such as this committee, and with others in the utility industry.

Today marks the 99th day since the spill occurred. We have made steady progress in the initial recovery work, including development of a Corrective Action Plan that includes comprehensive monitoring of the air, water and soil. It is important to note that according to the Tennessee Department of Health, the environmental monitoring analyzed to date has not shown any adverse health threat to the immediate or surrounding community, including air quality or drinking water supplies. On March 19, we began the initial phase of dredging ash from the Emory River channel adjacent to the failed storage facility. This activity is being thoroughly monitored and precautions are in place to prevent or minimize environmental impacts during the dredging process. The dredging plan was approved by the Tennessee Department of Environment and Conservation (TDEC) and the U.S. Environmental Protection Agency (EPA).

Chapter 5- On December 22, 2008, at 1:00 a.m., an ash disposal cell at the TVA Kingston Fossil Plant failed, causing the release of an estimated 5.4 million cubic yards of fly ash to the Emory and Clinch Rivers and surrounding areas. The release extended over approximately 300 acres outside the ash storage area. The failed cell was one of three cells at the facility used for settling the fly ash. The initial release of material created a wave of water and ash that destroyed three homes, disrupted electrical power, ruptured a natural gas line in a neighborhood located adjacent to the plant, covered a railway and roadways in the area, and necessitated the evacuation of a nearby neighborhood.

Shortly after learning of the release, EPA deployed an On-Scene Coordinator to the site of the TVA Kingston Fossil Plant coal ash release. EPA joined TVA, the Tennessee Department of Environment and Conservation (TDEC), the Roane County Emergency Management Agency, and the Tennessee Emergency Management Agency (TEMA) in a coordinated response (i.e., unified command in the National Incident Management System). EPA provided oversight, as well as technical advice, for the environmental response portion of TVA's activities. TVA has conducted extensive environmental sampling and shared results with EPA personnel. As discussed in more detail below, EPA staff and contractors have also conducted extensive independent sampling and monitoring to evaluate public health and environmental threats. In addition to providing information on environmental conditions at the site, EPA's data have also served as an independent verification of the validity of the TVA data.

Chapter 6- Coal combustion residuals (CCR) are one of the largest waste streams generated in the United States, with approximately 131 million tons generated in 2007. Of this, approximately 36% was disposed of in landfills, 21% was disposed of in surface impoundments, 38% was beneficially reused, and 5% was used as minefill. In comparison, EPA's Biennial Hazardous Waste Report shows that approximately 33.7 million tons of hazardous waste was generated in the United States in 2007. CCR typically contain a broad range of metals, including arsenic, selenium, and cadmium; however, the leach levels, using EPA's Toxicity Characteristic Leaching Procedure (TCLP), rarely reach the Resource and Conservation Recovery Act (RCRA) hazardous waste characteristic levels. Due to the mobility of metals and the large size of the typical disposal unit, metals (especially arsenic) may leach at levels of potential concern from impoundments and unlined landfills.

The beneficial use of CCR provides environmental benefits in terms of energy savings, greenhouse gas emission reductions, and resource conservation. In 2007, 56 million tons of CCR were reused. For example, use of CCR contributed to the construction of the Hoover Dam, the San Francisco-Oakland Bay Bridge, and the new I-35 bridge in Minneapolis, Minnesota. Many state environmental statutes and regulatory programs, as well as state road construction agencies, provide for the beneficial use of CCR. In 2007, use of coal fly ash as a substitute for Portland cement in concrete reduced energy use in concrete manufacturing by 73 trillion British thermal units (BTUs), with associated greenhouse gas emission reductions estimated at 12.5 million tons of carbon dioxide equivalent (MTCO2).

Chapter 7- Thank you, Mr. Chairman, for the opportunity to testify before the Subcommittee on Water Resources and Environment today. My name is Eric Schaeffer, and I am Director of the Environmental Integrity Project, a nonprofit and nonpartisan organization that advocates for more effective enforcement of federal environmental laws. I also served as director of the USEPA's civil enforcement program from 1997 to 2002. The testimony that follows is offered on behalf of myself and my colleague Lisa Evans, a senior attorney at Earthjustice and one of the nation's leading experts on coal ash. Our testimony will make the following points:

1) Coal ash is a hazardous material that tends to leak toxic metals into groundwater and surface water, especially when the ash is saturated or stored in wet ponds.
2) The discharge of wastewater from coal ash ponds, as well as the runoff from so-called dry landfills, can release arsenic, selenium and other pollutants in amounts known to be toxic to human health and aquatic life in our rivers and lakes. Despite the risks, discharges of toxic metals are generally not restricted under Clean Water Act permits at power plants and are often not even monitored.
3) Air pollution control equipment installed to comply with the Clean Air Act will generate thousands of tons of scrubber sludge at a typical power plant. USEPA and industry data show that the wastewater discharged from scrubber sludge treatment systems can release toxic metals like selenium in concentrations that are hundreds of times higher than water quality standards designed to protect aquatic life.
4) USEPA has promised to develop federal safeguards for the disposal of coal ash, but is also evaluating whether to set limits on the toxic discharges from ash and sludge treatment systems. The monitoring data indicate that such limits are overdue, and there is little time to lose.

Chapter 8- Madame Chairman, Members of the Committee and Distinguished Panelists:

My name is Dave Goss, former Executive Director of the American Coal Ash Association (ACAA) and I have been asked to appear before you today by ACAA's current Executive Director and its membership. ACAA promotes the recycling of coal combustion products (or CCPs) which include fly ash, bottom ash, boiler slag and air emission control residues, such as synthetic gypsum. It is our opinion, that the U.S. Environmental Protection Agency (EPA) regulatory determinations, made in 1993 and reaffirmed in 2000, are still correct that CCPs DO NOT warrant regulation as hazardous waste.

The recycling of these materials is a tremendous success story that has displaced more than 120 million tons of greenhouse gases since 2000. During that same period, more than 400 million tons of CCPs have been recycled in road construction, architectural applications, agriculture, mine reclamation, mineral fillers in paints and plastics, wallboard panel products, soil remediation and numerous other uses that would have required other materials if these CCP products were not available. Use of 400 million tons of CCPs displaces enough landfill capacity to equal 182 billion days of household trash.

The use of CCPs goes back more than forty years. In the last three decades, the EPA, other federal agencies, numerous universities and private research institutes have extensively studied CCP impact on the environment. The U.S. Department of Energy and the U.S. Department of Agriculture have both funded, conducted and evaluated mining and land case studies using a variety of applications. Consistently, these federal agencies found that when properly characterized, managed and placed, CCPs do not have a harmful impact on the environment or on public health.

EPA reported to Congress on March 31, 2009, results of data collected and analyzed by the Agency from the Tennessee Valley Authority ash spill on December 22, 2008. This data showed that there were no exceedances to drinking water or air quality standards. This information was based on hundreds of water samples and more than 26,000 air samples.

Chapter 9- Thank you for the opportunity to appear before the Subcommittee on Energy and Mineral Resources of the House Committee on Natural Resources. The subcommittee has called this hearing to address the question: "How Should the Federal Government Address the Health and Environmental Risks of Coal Combustion Waste?" Implicit in this question is the concern that coal combustion wastes may contain toxic constituents that pose long-term damage to water supplies and the resources that depend on them.

I have spent most of my professional career working on mining issues, with a particular emphasis on coal mining. I was also a member of the National Research Council (NRC) Committee that was called upon recently to study the disposal of coal combustion residues (CCRs) in coal mines as part of the mine reclamation process. That effort was especially relevant to the question posed by the committee.

I have two recommendations that respond to the question posed by the subcommittee. First and foremost, federal policy should treat the *disposal* of coal combustion residues — whether in coal mines, impoundments or landfills — as the option of last resort. Whenever possible, CCRs should be used for secondary beneficial purposes, and such use should be promoted through incentives for secondary use as well as disincentives for disposal. The NRC Committee recommended that secondary use of CCRs be "strongly encouraged." I would go further and argue that disposal of CCRs in coal mines, landfills, and impoundments should not be authorized unless and until the producer demonstrates a substantial and good faith effort to make the CCRs available for secondary use.

Chapter 10- Chairman Costa, and honorable members of the Committee, thank you for the opportunity to share Maryland's experience with coal combustion waste with you and, more importantly, for your interest in this very important issue.

We also greatly appreciate Congressman Sarbanes' interest and attention to issues surrounding the disposal of this by-product of producing energy from coal.

In 2006, the most recent year for which complete information is available from Maryland's Public Service Commission, coal generated 60.1% of the electricity generated in the State. In Maryland, there are five companies who generate coal combustion by-products at 9 facilities. Approximately 2 million tons of coal ash (fly and bottom ash) is generated annually from Maryland plants. Of that 2 million tons, approximately 1.6 million tons of coal ash is from the plants owned and operated by two companies, Constellation and Mirant.

In Maryland, the Maryland Healthy Air Act requires flue gas desulphurization equipment (known as "scrubbers") to be put in place by 2010 to reduce sulphur dioxide (SO2) emissions by 80%. A second phase of requirements in 2013 will increase the emission reductions to 85%. That equipment, while reducing SO2 emissions by over 200,000 tons will also increase the volume of scrubber sludge produced by 2.5 million tons. By 2013, therefore, facilities in Maryland will generate 4.5 million tons of CCWs.

As you are aware, coal combustion by-products are frequently reused. Currently, approximately 1 million tons, or one half of the coal ash produced annually, is beneficially used in Maryland. Fly ash can be reused for concrete manufacturing and in building material. It can also be used as structural fill in roadway embankments and development projects. (It can also be used in agricultural applications. While these are just a few of the reuse applications, there are many outstanding questions with regard to the safety of reuse.) For example, when used for structural fill, should liners be used; should there be defined distances between use of CCWs and potable water sources; should it be prohibited in shoreline areas such as the Chesapeake Bay Critical Area, source water protection areas, wetlands, or other areas of special concern; if used in agriculture, should it be applied to crops that are for human consumption. These are issues being examined as the State begins to develop a second phase of regulations to more effectively control reuse.

Chapter 11- Chairman Costa, and honorable members of the Committee, thank you for the opportunity to share Maryland's experience with coal combustion waste with you and, more importantly, for your interest in this very important issue.

We also greatly appreciate Congressman Sarbanes' interest and attention to issues surrounding the disposal of this by-product of producing energy from coal.

In 2006, the most recent year for which complete information is available from Maryland's Public Service Commission, coal generated 60.1% of the electricity generated in the State. In Maryland, there are five companies who generate coal combustion by-products at 9 facilities. Approximately 2 million tons of coal ash (fly and bottom ash) is generated annually from Maryland plants. Of that 2 million tons, approximately 1.6 million tons of coal ash is from the plants owned and operated by two companies, Constellation and Mirant.

In Maryland, the Maryland Healthy Air Act requires flue gas desulphurization equipment (known as "scrubbers") to be put in place by 2010 to reduce sulphur dioxide (SO2) emissions by 80%. A second phase of requirements in 2013 will increase the emission reductions to 85%. That equipment, while reducing SO2 emissions by over 200,000 tons will also increase the volume of scrubber sludge produced by 2.5 million tons. By 2013, therefore, facilities in Maryland will generate 4.5 million tons of CCWs.

As you are aware, coal combustion by-products are frequently reused. Currently, approximately 1 million tons, or one half of the coal ash produced annually, is beneficially used in Maryland. Fly ash can be reused for concrete manufacturing and in building material. It can also be used as structural fill in roadway embankments and development projects. (It can also be used in agricultural applications. While these are just a few of the reuse applications, there are many outstanding questions with regard to the safety of reuse.) For example, when used for structural fill, should liners be used; should there be defined distances between use of CCWs and potable water sources; should it be prohibited in shoreline areas such as the Chesapeake Bay Critical Area, source water protection areas, wetlands, or other areas of special concern; if used in agriculture, should it be applied to crops that are for human consumption. These are issues being examined as the State begins to develop a second phase of regulations to more effectively control reuse.

Chapter 12- Good morning, Mr. Chairman. My name is David Goss, Executive Director of the American Coal Ash Association. I sincerely appreciate the opportunity to address you, the members of the Committee and other distinguished experts appearing before you on this important topic. ACAA is an industry association of producers, marketers, end-users, researchers and others who support the beneficial use of what our industry refers to as coal combustion products, commonly known as CCPs. This includes coal ash and residues from air emission control systems such as synthetic gypsum products. These materials are the residuals from the burning of coal to generate electricity. By the very nature of the energy generation process utilizing coal, these byproducts cannot be eliminated entirely and must be managed like many other industrial byproduct streams. We consider CCPs to be mineral resources that if not used, become resources that are wasted.

In a perfect world, energy generation would not have any byproducts because the process would efficiently use all of the raw materials needed to generate electricity. Yet, the coal fueled generation process is not perfect. Even other energy options have consequential impacts, for example wind, which yields noise pollution and bird impingement. The coal-based energy generation industry generates byproducts including fly ash, bottom ash, slag and gypsum. The difference is that many of our products can replace or improve other commonly used commodities including portland cement and constituents which are used to produce concrete and other construction materials. The safe re-use of CCPs has a significant positive impact on this nation's mineral resources, its environment and economy. It is essential to promote and support activities that contribute to a more sustainable nation. By sustainable nation, I mean efficient, socially responsible and environmentally friendly usage of CCPs. I think the majority of us would agree that byproduct re-use which is environmental, health and safety conscience is much better than putting wastes in a disposal facility. Recognizing this common interest to promote safe and environmentally sound byproducts use, I am here to address how the beneficial use of CCPs contributes measurably to reduce environmental impact and is properly being regulated by the federal and state authorities.

Chapter 13- The question the subcommittee is exploring carries important, implicit understandings in its phrasing. There is implicit understanding that coal combustion waste (CCW) exists. There is implicit understanding that there are health and environmental risks with CCW. There is implicit understanding that the risks need be addressed. There is implicit understanding that federal action is needed to address the risks. I share the each of those understandings with the author(s) of the question, although I must admit resistance in reaching the last understanding.

My understandings are founded in 5½ decades of personal observation, management, and study of CCW. In the 1 950s I became responsible for removing, carrying, and dumping the "clinkers" from our coal furnace. They were put to "beneficial use," providing traction and filling ruts on the lane coming up the hill to the farmhouse. In the 1 960s, I became painfully aware that even beneficial use of these materials carries risks, as did everyone else who tried to skate on an icy road after the township trucks had spread cinders or who tripped on the cinder track during the hand-off in the mile relay. In the 1960s and 1970s, I was episodically subjected to the rain of fly ash and the taste and feel of sulfur dioxide in my throat when the wind was from the university's power plant in Champaign, Illinois. Since the mid-1980s, a significant portion of my professional career has been the study and evaluation of CCW, now remove from the air, and how best to manage it. My client base through the years has included individuals, coal companies, environmental organizations, power companies, governmental units, and citizens' groups.

My testimony today represents my personal understanding and opinions, and is not intended to represent those of any other individuals or organizations. My opinions and understanding have evolved and should continue to evolve as I learn more. If they don't, I should retire. I am not being paid to be here and my preparation for this hearing is similarly donated, although I am seeking reimbursement of direct travel expenses.

Chapter 14- I thank you for the opportunity to testify today concerning the health effects of exposure to coal combustion waste. I am Dr. Mary Fox, Assistant Professor in the Department of Health Policy and Management in the Johns Hopkins Bloomberg School of Public Health. I am a risk assessor with doctoral training in toxicology, epidemiology and environmental health policy. I am a core faculty member of the Hopkins Risk Sciences and Public Policy Institute where I teach the methods of quantitative risk assessment. In my research I evaluate the health risks of exposure to multiple chemical mixtures.

My testimony focuses on the health effects associated with exposure to coal combustion waste and assessing the public health risks of such exposures.

Chapter 15- Chairman Costa and Members of the Subcommittee, thank you for holding this hearing to consider the federal government's role in addressing the health and environmental risks of coal combustion waste. When mismanaged, coal combustion waste damages aquatic ecosystems, poisons drinking water and threatens the health of Americans nationwide. One of the dangers posed by coal combustion waste is disposal in coal mines, a practice that threatens the already heavily impacted communities and natural resources of our nation's coal mining regions.

I am Lisa Evans, an attorney for Earthjustice, a national non-profit, public interest law firm founded in 1971 as the Sierra Club Legal Defense Fund. Earthjustice represents, without charge, hundreds of public interest clients in order to reduce water and air pollution, prevent toxic contamination, safeguard public lands, and preserve endangered species. My area of expertise is hazardous and solid waste law. I have worked previously as an Assistant Regional Counsel for the Environmental Protection Agency enforcing federal hazardous waste law and providing oversight of state programs. I appreciate the opportunity to testify this morning.

The question before this subcommittee, how the federal government should address the risks of coal combustion waste, has a straightforward answer. Simply stated, the U.S. Environmental Protection Agency (EPA) must do what it committed to do in its final *Regulatory Determination on Wastes from the Combustion of Fossil Fuels*, published 8 years ago.[1] In that determination, mandated by Congress in 1980, EPA concluded that federal standards for the disposal of coal combustion waste under the Resource Conservation and Recovery Act (RCRA) and/or the Surface Mining Control and Reclamation Act (SMCRA) are required to protect health and the environment. EPA's commitment to set minimum federal disposal standards extended to coal ash disposed in landfills, lagoons and mines. Yet eight years later, and 25 years after Congress required this determination, EPA's commitment remains an entirely empty promise.

Chapter 16- "How should the Federal Government address the Health and Environmental Risks of Coal Combustion Waste"?

I live in a very conservative multi-cultural neighborhood that was once predominantly African American. Being an African American and having been exposed to the many facets of public service, I was soon able to transfer skill sets and assistance to this small community that was besieged by large corporations and landfill operators. For decades these corporations had targeted them with disposal of chemical waste and toxic materials. Too often, and on a continuing basis, large organizations and businesses too eager to turn a large profit margin, target communities of disproportionate underrepresented minority groups (i.e. African Americans, Alaska Natives, American Indians, Mexican Americans and Hispanic groups) for chemical and toxic waste disposal.

Often focusing on certain areas of disparity in subject matter areas such as education, criminal and environmental justice, these corporations prey on these groups' socioeconomic status to unfairly take advantage of their communities, homes and lifestyles. The impact of these criminal predators is long felt months if not years later when health issues arise, and property and home values diminish. State and County officials who often work hand in hand to appease these perpetrators have either left office or attribute their decisions to the greater good of county revenue generated from taxes, permits and fees imposed. The Maryland Department of Environment (MDE), an agency charged to protect the environment and public health of its citizens, has consistently failed the very citizens that have been aggrieved in the Evergreen Road and Waugh Chapel communities.

In: Coal Combustion Waste: Management and Beneficial Uses ISBN: 978-1-61728-962-0
Editor: Daniel D. Lowell

Chapter 1

MANAGING COAL COMBUSTION WASTE (CCW): ISSUES WITH DISPOSAL AND USE

Linda Luther

SUMMARY

In 2008, coal-fired power plants accounted for almost half of the United States' electric power, resulting in as much as 136 millions tons of coal combustion waste (CCW). On December 22, 2008, national attention was turned to issues regarding the waste when a breach in an impoundment pond at the Tennessee Valley Authority's (TVA's) Kingston, Tennessee, plant released 1.1 billion gallons of coal ash slurry. The estimated cleanup cost will likely reach $1.2 billion.

The characteristics of CCW vary, but it generally contains a range of heavy metals such as arsenic, beryllium, chromium, lead, and mercury. While the incident at Kingston drew national attention to the potential for a sudden catastrophic release of waste, the primary concern regarding the management of CCW usually relates to the potential for hazardous constituents to leach into surface or groundwater, and hence contaminate drinking water, surface water, or living organisms. The presence of hazardous constituents in the waste does not, by itself, mean that they will contaminate the surrounding air, ground, groundwater, or surface water. There are many complex physical and biogeochemical factors that influence the degree to which heavy metals can dissolve and migrate offsite—such as the mass of toxins in the waste and the degree to which water is able to flow through it. The Environmental Protection Agency (EPA) has determined that arsenic and lead and other carcinogens have leached into groundwater and exceeded safe limits when CCW is disposed of in unlined disposal units.

In addition to discussions regarding the potential harm to human health and the environment, the Kingston release brought attention to the fact that the management of CCW is essentially exempt from federal regulation. Instead, it is regulated in accordance with requirements established by individual states. State requirements generally apply to two broad categories of actions—the *disposal* of CCW (in landfills, surface impoundment, or mines) and its *beneficial use* (e.g., as a component in concrete, cement, or gypsum wallboard, or as structural or embankment fill).

In May 2000, partly as a result of inconsistencies in state requirements, EPA determined that national regulations regarding CCW disposal were needed. To date, regulations have not been proposed. However, on March 9, 2009, EPA stated that regulations to address CCW disposal in landfills and surface impoundments would be proposed by the end of 2009. Also, in March 2007, an advance notice of proposed rulemaking regarding the disposal of CCW in mines was released by the Department of the Interior's Office of Surface Mining (OSM). Draft rules have not yet been proposed. With regard to potential uses of CCW, EPA has stated that there have been few studies that would definitively prove that certain uses of CCW are safe, but that its use should include certain precautions to ensure adequate groundwater protection. It is unknown whether regulations regarding beneficial uses of CCW will be included in the upcoming rulemaking.

Some Members of Congress and other stakeholders have expressed concern regarding how CCW will ultimately be regulated. Among other issues, there is concern that the upcoming regulations will be either too far-reaching, and hence costly, or not far-reaching enough—meaning that they will not establish consistent, enforceable, minimal federal requirements applicable to CCW disposal units. On December 17, 2009, EPA issued a statement that its pending decision on regulating CCW would be delayed for a "short period due to the complexity of the analysis the agency is currently finishing."

Overview of Disposal and Use Issues

Coal fired power plants account for almost 45% of electric power generated in the United States. The coal combustion process at those facilities generates a tremendous amount of waste. In 2008, industry estimates indicate that 136 million tons of coal combustion waste (CCW) was generated.[1] That would make CCW the second largest waste stream in the United States, second to municipal solid waste, or common household garbage. How CCW is managed and how those management methods are regulated have come under increased scrutiny in the last year.

Coal combustion waste is managed in two ways: It may be *disposed of* in landfills or surface impoundment ponds, or in mines as minefill, or it may be *used* in some capacity (commonly referred to as "beneficial use")—for example, as a component in concrete, cement, or gypsum wallboard, or as structural or embankment fill. These management methods are largely unregulated at the federal level. Instead, they are regulated according to state requirements that vary from state to state.

On December 22, 2008, national attention was turned to potential risks associated with CCW management when a breach in an impoundment pond at the Tennessee Valley Authority's (TVA's) Kingston, Tennessee, plant released 1.1 billion gallons of coal fly ash slurry. The release covered more than 300 acres and damaged or destroyed homes and property. The sludge discharged into the nearby Emory and Clinch rivers, filling large areas of the rivers and resulting in fish kills. Sampling at the site in January 2009 found arsenic levels that exceeded the Environmental Protection Agency's (EPA's) Removal Action Level (contaminant levels at which time-critical response actions may be required).[2]

According to TVA, the estimated cleanup cost will likely reach $1.2 billion.[3] TVA recognized that this estimate could change significantly depending on the method of containment or the amount of ash ultimately disposed of as well as the impact of new coal ash laws and regulations that may be implemented at the state or federal level. Further, TVA's estimate does not include the potential costs associated with future regulatory actions, litigation, fines or penalties that may be assessed, final remediation activities, or other settlements. EPA and TVA have estimated that the cleanup may take two to three years.

In addition to the Kingston release, other events related to CCW management have attracted national media attention, as well as the attention of various stakeholders and some Members of Congress. For example, on December 30, 2008, a $54 million class action settlement was approved between Constellation Energy and Maryland residents after CCW that had been disposed of in a Gambrills, Maryland, quarry contaminated the owners' drinking water. Wells were determined to be contaminated with arsenic, lead, cadmium, and sulphates at levels above EPA drinking water standards.

Concern for a potential accidental release or contamination associated with CCW management is not new. However, the recent high-profile incidents have brought increased attention to the issue. Concerns about CCW management generally center around the following issues:

- The waste is generated in tremendous volumes and has been accumulating at some sites for decades. Individual power plants may generate thousands to hundreds of thousands of tons of the waste each year—the majority of which is disposed of onsite. Some plants have been in operation for decades (the site of the Kingston release has accumulated ash sludge since 1954), resulting in the disposal of millions of tons of CCW at individual plants across the United States.
- The waste likely contains certain hazardous constituents that EPA has determined pose a risk to human health and the environment. Those constituents include heavy metals such as arsenic, beryllium, boron, cadmium, chromium, lead, and mercury, and certain toxic organic materials such as dioxins and polycyclic aromatic hydrocarbon (PAH) compounds.
- Under certain conditions, hazardous constituents in CCW migrate and can contaminate groundwater or surface water, and hence living organisms. For example, EPA determined that the potential risk of human exposure to arsenic and other metals in CCW (via the groundwater-to-drinking-water pathway) increased significantly when CCW was disposed of in unlined landfills. That risk criterion was slightly higher for unlined surface impoundments.[4]
- According to EPA, the majority of new landfills and surface impoundments are constructed with liners and have groundwater monitoring systems. However, it is difficult to determine how many older units that may be operational do not have liners or groundwater monitoring.
- Although CCW contains hazardous constituents, it has been specifically exempt from federal hazardous waste management regulations. Instead, it is regulated in accordance with requirements established by the states. In 1999, EPA determined that national regulations regarding CCW disposal were needed, in part due to

inconsistencies in state requirements. Since then, various surveys have been conducted and data gathered, but EPA has not proposed regulations.

Members of the public, particularly those near utility plants, have expressed concern that their health or property values may be affected by either a sudden release of waste, as in Kingston, or the gradual release of contaminants. Industry organizations insist that the waste is generally safe and does not pose a significant risk that would warrant the increased cost of more stringent management—costs that would be ultimately borne by rate payers. They also argue that being required to manage the waste according to hazardous waste regulations would limit its potential for use—thereby increasing the amount that must be disposed of. Environmental organizations argue that the Kingston spill was a warning sign of spills to come and that there is currently inadequate oversight and monitoring of either existing or closed disposal sites.

Some Members of Congress have also expressed concern over these issues, both before and after the Kingston release. Like other stakeholders, their concerns have stretched across various areas including the role that coal mining plays in our economy, the role that coal-fired utilities play as a major source of domestic energy, the federal role in the regulation of CCW, as well as the potential risks posed to their constituents if CCW is managed improperly.[5]

EPA has been studying how best to regulate CCW since at least 1980. Waste management is regulated under provisions of the Resource Conservation and Recovery Act (RCRA, 42 U.S.C. §690 1 *et seq.*). In May 2000, EPA determined that CCW did not warrant regulation under subtitle C of RCRA (the federal *hazardous* waste requirements). EPA did, however, determine that national regulations under subtitle D of RCRA (the *solid* waste management requirements) were warranted for CCW when it is disposed in landfills or surface impoundments. EPA also found that CCW used to fill surface or underground mines warranted regulation under RCRA's solid waste requirements or possibly under modifications to existing regulations established under authority of the Surface Mining Control and Reclamation Act (SMCRA).[6] Since then, various surveys have been conducted, reports issued, and data gathered, but no federal regulations have been proposed.

On March 9, 2009, in the wake of the Kingston release, EPA declared its intent to move forward with CCW regulations to address the management of coal combustion residuals. EPA stated that regulations would be proposed for public comment by the end of 2009.

Since EPA's statement, industry and environmental groups, state government representatives, and some Members of Congress have expressed concerns regarding how the waste will ultimately be regulated. Generally those concerns center around whether CCW will be regulated as a solid or hazardous waste. Subsequently, representatives with EPA have declared that, although they felt that regulating CCW under RCRA's solid waste management requirements (e.g., subtitle D's landfill criteria and permitting requirements) would provide sufficient protection, the agency had no authority to do so. Instead, their only existing authority was to regulate the waste under RCRA's hazardous waste management requirements (for more information, see "The Current Rulemaking"). On December 17, 2009, EPA issued a statement that its pending decision on regulating CCW would be delayed for a "short period due to the complexity of the analysis the agency is currently finishing."[7]

Regardless of the ultimate choice of waste management, the amounts of CCW generated each year are tremendous. As power plant emission standards become more stringent and air emission control devices capture more contaminants, both the total waste generated and the amount of toxins in the waste can be expected to increase.

Table 1. Types of Coal Combustion Waste

Waste Type	Description	Percentage of Total Generated
Fly Ash	A product of burning finely ground coal in a boiler to produce electricity. It is generally captured in the plant's chimney or stack through a particulate control device (e.g., electrostatic precipitators or fabric filters). It consists mostly of silt-sized and clay-sized glassy spheres, giving it a consistency somewhat like talcum powder.	57%
Flue Gas Desulfuri-zation (FGD) Material	Flue gas desulfurization (FGD) is a chemical process implemented in order to meet emission requirements in the Clean Air Act applicable to sulfur dioxide (an emission associated with acid rain). The goal of the process is to chemically combine the sulfur gases released in coal combustion by reacting them with a sorbent, such as limestone (calcium carbonate), lime (calcium oxide), or ammonia. Depending on the FGD process used at the plant, the material may be a wet sludge or a dry powder. The wet sludge is likely predominantly calcium sulfite or calcium sulfate. The dry material generally consists of a mixture of sulfites and sulfates.	24%
Bottom Ash	A coarse, gritty material, these agglomerated ash particles are those that are too large to be carried in flue gases. They impinge on the furnace walls or fall through open grates to an ash hopper at the bottom of the furnace. The material is taken from the bottom of the boiler furnace either in its dry form or as a slurry (via the addition of water). It has a porous surface structure and is coarse, with grain sizes spanning from fine sand to fine gravel.	17%
Boiler Slag	This type of ash collects at the base of certain furnaces that are quenched with water. When molten slag comes in contact with quenching water, it fractures, crystallizes, and forms pellets. This boiler slag material is made up of hard, black, angular particles that have a smooth, glassy appearance. The particles are uniform in size, hard, and durable, with a resistance to surface wear.	<2%

Source: Table generated by the Congressional Research Service (CRS) using information from the Environmental Protection Agency's "Wastes - Resource Conservation - Reduce, Reuse, Recycle - Industrial Materials Recycling" Web page regarding Coal Combustion Products, at http://www.epa.gov/osw/conserve/rrr/ imr/ccps/index.htm and the U.S. Geological Survey, Fact Sheet 076-01, "Coal Combustion Products," available at http://pubs.usgs.gov/fs/fs076-01/fs076-01.html.

Notes: The approximate percentage of total CCW generated was determined using data from American Coal Ash Association (ACAA), "2007 Coal Combustion Product (CCP) Production & Use Survey Results (Revised)," available at http://www.acaa-usa.org/displaycommon.cfm? an=1&sub articlenbr=3.

To provide information and context on this issue, this chapter discusses the nature of the waste itself; potential risks associated with its management; the regulatory history of CCW management requirements, including why CCW is exempt from federal regulations and issues associated with the current rulemaking; and CCW management options (e.g., landfill and surface impoundment disposal) and the likely state and federal requirements associated with those management methods.

This chapter does not provide risk analysis regarding the disposal or use of CCW, or information regarding the potential fate and transport of hazardous constituents. It also does not discuss details regarding the Kingston release, such as determinations regarding the cause of the release or details of the cleanup. Numerous studies on those topics have been conducted, findings from which are cited or summarized where appropriate. This chapter focuses primarily on the issues associated with CCW disposal in landfills and surface impoundments, but also provides summary information on issues associated with its disposal in mines and its "beneficial use."[8]

THE NATURE OF COAL COMBUSTION WASTE

Each step of the coal combustion process results in different types of waste. The characteristics and potential risks associated with that waste vary according to many factors. To minimize the potential negative impacts associated with CCW management, it is necessary to understand the characteristics of the specific type of CCW being handled as well as the physical environment in which it is placed.

Types of Coal Combustion Waste

Coal combustion waste consists of inorganic residues that remain after pulverized coal is burned. At various stages of the coal combustion process, different types are generated. These residues include both coarse particles that settle to the bottom of the combustion chamber and fine particles that are removed from the flue gas by electrostatic precipitators, scrubbers, or fabric filters. Factors such as the source of the coal burned at a plant and the technology used (both to burn the coal and to filter the ash) have bearing on CCW's characteristics and potential toxicity. Table 1 describes the different types of CCW generated.

Waste Characteristics

The physical and chemical characteristics of each type of CCW have bearing on both its potential for use (e.g., as a component in concrete or gypsum wallboard) and its potential to present some level of risk to human health or the environment. In 2006, a study by the National Research Council (NRC) identified several factors that influence the physical and chemical characteristics of CCW.[9] Included among the factors are:

- **The chemical characteristics of the source coal.** The waste itself represents noncombustible constituents in coal. Therefore, its characteristics are strongly influenced by the source coal itself (e.g. lignite, bituminous).
- **The chemical characteristics of any co-fired materials.** Some coal-fired boilers, especially at non-utilities (e.g., boilers at industrial, commercial, or chemical facilities), may be co-fired with materials such as wood, biomass, plastics, petroleum coke, tire-derived fuel, refuse-derived fuel, or manufactured gas plant wastes.
- **The processes or technology used at individual utility plants.** Waste characteristics are affected by the particular combustion technology, air emission control devices used to capture regulated contaminants (e.g., sulfur dioxide, nitrogen oxide, mercury), and residue-handling technology (collection systems result in either dry or wet residues) used at the plant.

While the components of each type of ash vary, depending on these factors, all CCW will likely include certain amounts of toxic constituents, primarily heavy metals such as arsenic, beryllium, boron, cadmium, chromium, cobalt, lead, manganese, mercury, molybdenum, selenium, strontium, thallium, and vanadium. The waste will also likely include a certain level of toxic organic materials such as dioxins and polycyclic aromatic hydrocarbon (PAH) compounds.

With regard to the source coal, the U.S Geological Survey (USGS) maintains a database of coal quality characteristics of coal basins in the United States.[10] The three types of coal most often used in utility boilers, bituminous, subbituminous, and lignite, vary in terms of their chemical composition, ash content, and geological origin. Some of the principal components in the fly ash from these types of coal are silica, alumina, iron oxide, potassium, calcium, and magnesium. These same components can make CCW usable as an ingredient in Portland cement or as a soil amendment.

Knowledge of coal chemistry—as well as the technology used to fire, filter, and collect it—is important to determine which mitigation procedure will be most efficient in reducing the amount of hazardous material potentially generated as waste.

Potential Risks Associated with CCW Management

Generally, there are two potential risks associated with the disposal or use of CCW. Disposal or use that involves direct applications of the waste to the ground may allow hazardous constituents in the waste to leach from the material, migrate, and contaminate groundwater or surface water and, ultimately, living organisms. Also, the land disposal of high volumes of liquid waste could result in a sudden release, as occurred at Kingston.

Surface or Groundwater Contamination

Although the possibility exists that hazardous constituents in CCW could become airborne, the primary concern regarding the management of the waste usually relates to the potential for hazardous constituents to leach into surface or groundwater, and hence contaminate drinking water, surface water, or biota. The presence of hazardous constituents in the waste does not, by itself, mean that they will contaminate the surrounding air, ground, groundwater, or surface water. The 2006 NRC report stated that there are many complex physical and biogeochemical factors that influence the degree to which heavy metals can essentially dissolve and migrate offsite. Those factors include:[11]

- **The volume and degree to which water is able to flow through the waste.** Water, such as precipitation or groundwater flow, is the primary mechanism for the transport of hazardous constituents through the waste. EPA has found that contaminants have a significantly higher likelihood of migrating away from the disposal site (i.e., landfill or surface impoundment) if they are disposed of in an unlined disposal unit (i.e., a scenario in which water is able to flow through the waste).
- **The chemistry, particularly the pH, of the water that flows through or contacts the waste.** Different metals commonly found in CCW are soluble in acidic (high pH), alkaline (low pH), or neutral environments. Many types of CCW are themselves alkaline and capable of neutralizing acidity. This is one reason why certain types of CCW are placed in mines to treat acid mine drainage.
- **The leachable mass of toxic constituents present in the waste.**

While general factors that contribute to contamination migration are known, it is difficult to determine the degree to which actual or potential contamination is being monitored at CCW disposal sites or sites where it has been placed directly on soil (e.g., as structural or embankment fill). EPA has documented selected cases of damages associated with disposal in landfills and surface impoundments.[12] However, there is little data regarding contamination associated with its disposal in mines or the "beneficial use" of CCW when it is place directly on land—such as when used as embankment or structural fill. That does not mean that contamination *has* occurred in those uses—only that it is known that CCW has been managed in a way that contamination could be anticipated. The degree to which such data may be tracked in individual states is difficult to determine.

A Sudden of Release of Liquid Waste

Surface impoundment ponds hold liquid waste that has been sluiced from the power plant to the disposal area. It is generally held within the pond by depositing it in a natural depression in the ground or through the use of a dike of some sort (see discussion regarding the use and regulation of surface impoundments in "Landfill and Surface Impoundment Disposal"). The Kingston release resulted from a rupture in an impoundment dike.

On March 9, 2009, in an attempt to avoid catastrophic releases such as that in Kingston, EPA sent a request for information to the owners and operators of CCW impoundment units.[13] The information-gathering was intended to assist in prioritizing surface impoundment ponds for inspection, and, ultimately, to assess the structural integrity of the units. EPA regional offices were asked to assist in identifying facilities that they considered priorities.

EPA sent the assessment survey to 61 utility headquarters and 162 individual facilities, requesting information regarding, among other factors, the stability of liquid-holding surface impoundment units. Responses to EPA's survey request identified 584 impoundment units (almost 300 more than EPA originally thought were operating). Survey responses also indicated that 49 units at 30 different locations were deemed "high hazard units." Such a rating is not an indication of the structural integrity of a unit or an assessment of its potential for failure. Rather, the rating allows dam safety and other officials to determine where significant damage or loss of life may occur if there is a structural failure of the unit. [14] EPA's intent in determining the dam safety rating was to assist in prioritizing inspections at individual facilities.

REGULATORY HISTORY AND CURRENT RULEMAKING

The Resource Conservation and Recovery Act (RCRA, 42 U.S.C. §6901 *et seq.*) [15] provides the general guidelines under which all waste is managed. It also includes a congressional mandate to EPA to develop a comprehensive set of regulations to implement the law (also commonly referred to as RCRA). Enacted in 1976, RCRA was intended, in part, to protect human health and the environment from the potential hazards of waste disposal and to ensure that wastes are managed in an environmentally sound manner.

The evolution of CCW regulation involves a long and somewhat complicated history. To understand issues associated with CCW disposal regulations and the current rulemaking process, it is useful to understand EPA's current authority under RCRA to regulate solid and hazardous waste; the terms of the "Bevill amendment," which excluded CCW from regulation under RCRA's hazardous waste requirements; EPA's actions in response to Bevill amendment directives; and issues associated with the current rulemaking process—particularly questions regarding EPA's potential to regulate CCW as solid or hazardous waste.

Waste Management Requirements Potentially Related to CCW

Broadly, industrial waste is regulated pursuant to standards applicable to "solid waste" and "hazardous waste." The current debate regarding CCW management centers around determining under which of those categories CCW belongs. To understand some of the challenges associated with the current rulemaking, it is useful to understand how a waste is identified as a solid waste or a hazardous waste (under the regulatory definition), and EPA's current authority to regulate each category of waste.

Identifying Solid and Hazardous Waste

RCRA regulations define solid waste broadly as any discarded material.[16] The regulations specify that a *solid* waste becomes a *hazardous* waste[17] by exhibiting one or more of the following characteristics—toxicity, reactivity, ignitability, or corrosivity. If CCW were to be characterized as hazardous, it would likely be because hazardous constituents in the waste exceed regulatory *toxicity* levels. EPA requires that toxicity characteristics be determined using the Toxicity Characteristic Leaching Procedure (TCLP).[18] The TCLP test is intended to simulate conditions that would likely occur in a landfill, and measures the potential for toxic constituents to seep or "leach" into groundwater.

Generally, CCW does not "fail" TCLP (that is, a given sample of waste generally does not exceed toxicity levels for certain contaminants like lead, arsenic, selenium, or other heavy metals). However, the nature of CCW is unique, relative to other hazardous waste. That is, it is generated in huge volumes, with large amounts of inert, benign materials that effectively dilute what may be an overall significant amount of hazardous constituents in an entire CCW landfill or surface impoundment.

Another means by which a waste may be identified as hazardous is by EPA specifically listing it as such (hence commonly referred to as "listed wastes").[19] One category of listed waste is "source specific waste." This list of waste includes wastes from specific industries, such as petroleum refining or pesticide manufacturing. If CCW were determined to be a hazardous waste it would likely be a specifically listed waste.

In addition to specifically listing wastes as hazardous, certain wastes may be specifically *excluded* from the definition of hazardous waste or solid waste. Materials may be excluded for various reasons, including public policy, economic impacts, prior regulation, lack of data, or the waste's high volume and low toxicity. The decision to exclude these materials from the solid waste definition is a result of either congressional action (embodied in the statute) or EPA policy making (embodied in the regulations). For example, Congress excluded CCW from the definition of hazardous waste pending additional study from EPA (see discussion below regarding "CCW's Regulatory Exemption Under "the Bevill Amendment"").[20]

Solid Waste Management Requirements

Subtitle D of RCRA establishes state and local governments as the primary planning, regulating, and implementing entities for the management of non-hazardous solid waste, such as household garbage and non-hazardous industrial solid waste. RCRA specifically requires EPA to regulate solid waste management facilities that accept household hazardous waste or hazardous waste from "small quantity generators." Such a specific directive indicates that EPA *does not* have the authority to regulate other types of disposal facilities, such as those that receive CCW.

Hazardous Waste Management Standards

Subtitle C of RCRA created a hazardous waste management program that, among other elements, directed EPA to develop certain waste management criteria. Under subtitle C, EPA is authorized to establish a system for controlling hazardous waste from the time it is generated until its ultimate disposal (i.e., from "cradle to grave").

Under subtitle C, hazardous waste treatment, storage, and disposal facilities (TSDFs) are required to have permits, to comply with operating standards specified in the permit, to meet financial requirements in case of accidents, and to close their facilities in accordance with EPA regulations. The 1984 amendments imposed a number of new requirements on TSDFs with the intent of minimizing land disposal. Bulk hazardous liquid wastes are prohibited from disposal in any landfill, and severe restrictions are placed on the disposal of containerized hazardous liquids, as well as on the disposal of nonhazardous liquids in hazardous waste landfills. EPA was directed to review all wastes that it defined as hazardous and to make a determination as to the appropriateness of land disposal for them. Minimum technological standards were set for new landfills and surface impoundments, requiring, in general, double liners, a leachate collection system, and groundwater monitoring.

As required under subtitle C, EPA proposed hazardous waste management regulations in 1978.[21] In these proposed regulations, EPA identified six categories of wastes it deemed "special wastes" (including fossil fuel combustion wastes) which would be deferred from hazardous waste management requirements until the completion of further study and assessment to determine their risk to human health and the environment. These special wastes were identified because they typically were generated in large volumes and, at the time, were believed to pose less of a risk to human health and the environment than wastes identified for regulation as hazardous waste.

In 1980, the Solid Waste Disposal Act Amendments of 1980 amended RCRA in several ways, including exempting "special wastes" from regulation under subtitle C until further study and assessment of risk could be performed. This section of the law is frequently referred to as the "Bevill Amendments."

Authority to Address an Imminent and Substantial Endangerment

Section 7003 of RCRA (42 U.S.C. § 6973) provides EPA with broad enforcement tools that can be used to abate conditions that may present an imminent and substantial endangerment to health or the environment. Section 7003 allows EPA to address situations where the handling, storage, treatment, transportation, or disposal of any solid or hazardous waste may present such an endangerment. In these situations, EPA can initiate judicial action or issue an administrative order to any person who has contributed or is contributing to such handling, storage, treatment, transportation, or disposal to require the person to refrain from those activities or to take any necessary action.

Section 7003 is available for use in several situations where other enforcement tools may not be available. For example, Section 7003 can be used at sites and facilities that are not subject to subtitle C of RCRA or any other environmental regulation (as may be the case at CCW disposal or use sites).

Action under Section 7003 may be initiated if the following three conditions are met:

1. Conditions may present an imminent and substantial endangerment to health or the environment—such conditions generally require careful documentation and scientific evidence. However, the endangerment standard under RCRA has generally been broadly interpreted.
2. The potential endangerment stems from the past or present handling, storage, treatment, transportation, or disposal of any solid or hazardous waste.
3. The person has contributed or is contributing to such handling, storage, treatment, transportation, or disposal.[22]

Under Section 7003, EPA may take action as deemed necessary, determined on a case-by-case basis. Further, it gives EPA authority to obtain relevant information regarding potential endangerments.

Section 7003 authority has been cited by some industry representatives as one alternative to EPA to regulate CCW management. Proponents argue that such an approach would allow EPA to enforce disposal practices or uses that pose a potential threat. Opponents of this approach argue that it is a resource-intensive method of enforcement—one that would require

EPA to gather substantial amounts of information on individual disposal sites, as opposed to implementing a consistent national approach to regulation.

CCW's Regulatory Exemption under "the Bevill Amendment"

In the months before hazardous waste regulations were finalized in 1980, Congress debated RCRA reauthorization. In February 1980, Representative Tom Bevill introduced an amendment to the Solid Waste Disposal Act Amendments that would require EPA to defer the imposition of hazardous waste regulatory requirements for fossil fuel combustion waste and discarded mining waste until data regarding the materials' potential hazard to human health or the environment could be analyzed. Congressman Bevill stated that EPA's intent to regulate such waste as hazardous would discourage the use of coal and constitute an unnecessary burden on the utility industry.[23] In anticipation of the enactment of this legislation, according to EPA, the agency excluded the regulation of fossil fuel combustion waste from its final hazardous waste regulations.[24]

P.L. 96-482, the Solid Waste Disposal Act Amendments of 1980, was enacted on October 12 of that year. The law was intended, in part, to provide EPA with stronger enforcement authority to address illegal dumping of hazardous waste. The final version included Representative Bevill's amendment, which excluded the following large-volume wastes from the definition of hazardous waste under Subtitle C of RCRA:

- Waste generated primarily from the combustion of coal (e.g., fly ash waste, bottom ash waste, slag waste, and flue gas emission control waste) or other fossil fuels.
- Solid waste from the extraction, beneficiation, and processing of ores and minerals, including phosphate rock and overburden from the mining of uranium ore.
- Cement kiln dust waste.

The Bevill amendment specified that the hazardous waste exclusion would be held pending completion of a study and report to Congress by EPA for each waste category. Factors to be addressed in each study were specified under Section 8002 of RCRA. For example, EPA was required to determine the potential danger, if any, posed by each form of waste to human health or the environment; identify documented cases in which danger to human health or the environment had been proved; identify then-current disposal practices, alternatives to those disposal methods, and the costs of such alternatives; and identify then-current uses and potential future uses of coal combustion products.[25] Within six months of each report to Congress, EPA was directed to make a regulatory determination regarding whether the waste in question warranted regulation as a hazardous waste under Subtitle C of RCRA.

EPA Actions from the Bevill Amendment to Kingston

Since 1980, EPA has conducted various studies, submitted reports to Congress, and made regulatory determinations in response to the directives in the Bevill amendment. Those actions primarily address issues associated with landfill and surface impoundment disposal. EPA has also conducted studies into the beneficial use of CCW. In addition, both EPA and the Department of the Interior's Office of Surface Mining (OSM) have conducted various activities related to mine placement of CCW. Selected actions undertaken by both EPA and OSM are summarized in Table 2.

Table 2. EPA and OSM Actions in Response to Bevill Amendment Requirements Selected Actions and Findings Applicable to Coal Combustion Waste

Date	Document Type	Summary of Findings
Oct. 31, 1982	Deadline	EPA missed its statutory deadline for submitting its fossil fuel combustion waste report to Congress.
Feb. 1988	Report to Congress (RTC)	EPA published its "Report to Congress on Wastes from Combustion of Coal by Electric Utility Power Plants." The RTC found that the four large volume waste streams studied (fly ash, bottom ash, boiler slag, and flue gas emission control waste) were not a major concern. Trace constituents in the wastes, including arsenic, barium, cadmium, chromium, lead, mercury, and selenium, may present risks to human health and the environment. However, the data also indicated that these wastes generally do not exhibit RCRA hazardous waste characteristics. Further, the RTC concluded that current waste management practices appear to be adequate. The RTC also indicated that as of 1988, coal-fired electric utilities spent about $800 million per year for CCW disposal, and that costs would increase to $3.7 billion per year if CCW was regulated as hazardous waste under RCRA's Subtitle C. This chapter addressed wastes generated from the combustion of coal by electric utility power plants, but did not address co-managed wastes (independent power producing facilities that are co-managed with certain other CCW), other fossil fuel combustion wastes, and wastes from non-utility boilers. Those "remaining wastes" were addressed in a subsequent RTC in 1999.
Aug. 31, 1988	Deadline	EPA missed its statutory deadline for making a regulatory determination regarding wastes studied in its February 1988 RTC.
Aug. 9, 1993	Regulatory determination (58 FR 42466)	EPA concluded that the four waste streams studied in the 1988 RTC did not warrant regulation as hazardous waste under Subtitle C of RCRA. EPA determined that it required more time to research the "remaining wastes" to make an appropriate determination.
March 31, 1999	Report to Congress	EPA published its "Report to Congress on Wastes from the Combustion of Fossil Fuels." This chapter addresses "remaining wastes" identified in the 1988 RTC and 1993 regulatory determination.
May 22, 2000	Regulatory determination (65 FR 32214)	This determination applied to large-volume CCW generated at electric utility and independent power producing facilities, and non-utilities. EPA concluded that these wastes did not warrant regulation under subtitle C of RCRA, but that national regulations under RCRA's subtitle D (solid waste requirements) were warranted for CCW when it is disposed of in landfills or surface impoundments. Further, to consistently regulate such waste across all waste management scenarios, the agency stated its intent to

Table 2. (Continued)

Date	Document Type	Summary of Findings
		promulgate such national requirements. The agency also concluded that no additional regulations were warranted for CCW used beneficially. EPA stated that the agency did not wish to place any unnecessary barriers on the beneficial use of fossil fuel combustion wastes so that they can be used in applications that conserve natural resources and reduce disposal costs.
May 2001 - May 2004	Public Meetings	EPA held several meetings with stakeholders regarding the use and disposal of coal combustion byproducts. In the public notice for its March 2004 meetings, EPA stated that the "Agency remains concerned about coal combustion byproducts because of the potential for environmental damage; the lack of ground-water protection via monitoring and/or liners; and widely varying state regulatory programs."
Dec. 2002	Reports	EPA issued two draft reports "Regulation and Policy Concerning Mine Placement of Coal Combustion Waste in Selected States" and "Mine Placement of CCW: State Program Elements Analysis" (final versions do not appear to have been released). The reports review and summarize current state regulations and policies concerning the placement of CCW in surface and underground mines.
March 1, 2006	Report	The National Academy of Sciences' National Research Council (NRC) issued a report, *Managing Coal Combustion Residues in Mines*. Among other recommendations, NAS recommends that the Department of the Interior's Office of Surface Mining (OSM) take the lead in CCW disposal standards under the Surface Mining Control and Reclamation Act of 1977 (SMCRA, the primary federal law that regulates the environmental effects of coal mining). EPA is working with OSM as they amend the SMCRA regulations to better address minefilling in active coal mines as well as federally funded abandoned mines.
August 2006	Report	EPA and the U.S. Department of Energy issued a joint report *Coal Combustion Waste Management at Landfills and Surface Impoundments*, 1994-2004. The report evaluated CCW disposal practices and state regulatory requirements at landfills and surface impoundments that were permitted, built, or laterally expanded between January 1, 1994, and December 31, 2004. In part, the report concluded that, since the 1988 RTC, a majority of the states reviewed for the study tightened regulation of landfill liners, leachate-collection systems, and groundwater monitoring for new disposal units.
March 6, 2007	Advanced Notice of Proposed Rulemaking (72 FR 12025)	In response to the March 2006 NRC report on managing CCW in coal mines, OSM released an ANPR regarding "Placement of Coal Combustion Byproducts in Active and Abandoned Coal Mines." The ANPR cites various findings and recommendations in the NRC report as the basis for the initiation of the rulemaking process.
July 9, 2007	Report	EPA's Office of Solid Waste issued "Coal Combustion Waste Damage Case Assessment." In that report, EPA determined that there have been 24 cases of proven damage and 43 cases of potential damage associated with CCW landfills and surface

Table 2. (Continued)

Date	Document Type	Summary of Findings
		impoundments. (Cases of alleged damage were submitted for review to EPA by environmental organizations. EPA also collected information from its own experience and from state agencies.) In each case there has been either proven damage to surface water or to groundwater. In some cases, elevated levels of polychlorinated biphenyls (PCBs), chromium, arsenic, cadmium, nickel, beryllium, selenium, iron, and other metals were found. Potential impacts to human health and the environment that were observed included contaminated well water and fish-kills.
Aug 6, 2007	Report	EPA's Office of Solid Waste issued "Human and Ecological Risk Assessment of Coal Combustion Wastes." The draft risk assessment conducted by EPA sought to quantify human health and ecological risks associated with current disposal practices for high-volume CCW in landfills and surface impoundments. In part, the risk assessment stated that risks from clay-lined liners are lower than unlined units, but that risks were still well above risk criteria for arsenic and thallium for landfills and arsenic, boron, and molybdenum for surface impoundments. Composite liners effectively reduce the risks from all constituents below the risk criteria for both landfills and surface impoundments. Further, although it is likely that new landfills will have some type of liner, it is not known how many unlined units continue to operate in the United States.
Aug. 29, 2007	Notice of Data Availability (NODA)	EPA issued a Notice of Data Availability on "Disposal of Coal Combustion Wastes in Landfills and Surface Impoundments." Documents made available under the NODA were the August 2006 joint EPA/DOE report on report on CCW Management at Landfills and Surface Impoundments; EPA's Aug. 2007 risk assessment; and EPA's July 2007 damage case assessments. EPA made these documents available and sought public comments on how, if at all, the information should affect EPA's decisions as it continued to follow up on its Regulatory Determination for CCW disposed of in landfills and surface impoundments. EPA stated that it would "consider all the information provided through the NODA, the comments and new information submitted on the NODA, as well as the results of a subsequent peer review of the risk assessment, as it continued to follow up on its Regulatory Determination for CCW disposed of in landfills and surface impoundments."
Feb. 5, 2008	NODA	The comment period related to the August 2007 NODA was extended until February 11, 2008.
Feb. 12, 2008	Report	EPA Office of Solid Waste and Emergency Response, Economics, Methods, and Risk Analysis Division issued "Waste and Materials-Flow Benchmark Sector Report: Beneficial Use of Secondary Materials-Coal Combustion Products." Among other information, the report provides an overview of key beneficial uses of CCW and an analysis of its benefits and potential impacts.

Source: This table was prepared by CRS based on a review of the public record. In particular, EPA's "Fossil Fuel Combustion (FFC) Waste Legislative and Regulatory Time Line," available at http://www.epa.gov/osw/nonhaz/ industrial/special/fossil/regs.htm.

As discussed above, EPA first stated its intent to develop regulations applicable to CCW management in May 2000. In its regulatory determination, EPA concluded that CCW did not warrant regulation as hazardous waste pursuant to provisions of subtitle C. However, EPA stated that it was convinced that CCW could pose risks to human health and the environment if not properly managed, and there is sufficient evidence that adequate controls may not be in place. The agency cited, for example, that most states can require newer units to include liners and groundwater monitoring, but that 62% of existing utility surface impoundments do not have groundwater monitoring. Further, EPA stated:

> ... [I]n light of the evidence of actual and potential environmental releases of metals from these wastes; the large volume of wastes generated from coal combustion; the proportion of existing and even newer units that do not currently have basic controls in place; and the presence of hazardous constituents in these wastes; we believe, on balance, that the best means of ensuring that adequate controls are imposed where needed is to develop national subtitle D regulations.[26]

EPA stated its decision to establish national regulations under RCRA subtitle D for CCW that is disposed of in landfills or surface impoundments or used to fill surface or underground mines. Since that decision, EPA gathered data, issued various reports, and held public hearings, but did not propose regulations.

The Current Rulemaking

In the wake of the Kingston release, EPA again stated its intent to develop regulations applicable to CCW disposal. On March 9, 2009, EPA announced that it was moving forward on developing regulations to address the management of CCW. The agency stated that it planned to propose regulations by the end of 2009.

In its May 2000 regulatory determination, EPA stated that it would establish national regulations under subtitle D of RCRA, and specifically sited sections 1008(a) and 4004(a)[27] of the law as the basis of its authority to regulate for CCW disposed in landfills or surface impoundments or used to fill surface or underground mines. However, in 2009, statements attributed to an EPA representative indicated that the authority previously cited was not sufficient to regulate CCW under subtitle D.[28] Instead, subtitle D gave EPA only the authority to regulate sanitary landfills (discussed above). Specifically, under subtitle D, Congress requires the "upgrading of open dumps" (42 U.S.C. § 6945) and directs EPA to determine the adequacy of certain guidelines and criteria (42 U.S.C. § 6949a) applicable to certain solid waste management and disposal facilities. Those sections of RCRA apply specifically to facilities that may receive hazardous household wastes or hazardous wastes from small-quantity generators.

Solid waste facilities that receive CCW are not specifically identified in the law. Therefore, according to EPA, the agency is not authorized to promulgate enforceable regulations under subtitle D (e.g., to establish landfill criteria or require states to include CCW disposal units in their solid waste permitting programs.

Table 3. Summary of Potential Regulatory Authorities to Address CCW Management Selected Stakeholder Arguments "For" and "Against" Potential Regulatory Approaches

Statutory Authority	Description	Proponent Arguments "For"	Opponent Arguments "Against"
RCRA Subtitle C- Hazardous Waste Management Requirements	CCW would be regulated as a hazardous waste. Land disposal of the waste would be prohibited unless certain criteria were met, such as landfills and surface impoundments would be required to have synthetic liners and groundwater monitoring. Further, states would be obliged to apply for federally enforceable permits from EPA. Under this approa-ch, EPA has the option to exclude CCW from the definition of hazardous waste under certain conditions, such as when recycled under specific conditions.	Environmental groups favor this approach because they argue that it is the only option that would ensure that consistent, enforceable, minimum federal standards would be applied to CCW disposal units and its uses. It would also provide EPA with enforcement and inspection authority that is currently lacking.	Industry groups and state agencies argue that this approach is too strict, would result in significant logistical challenges, and would cost billions of dollars. Some industry groups also believe that a hazardous waste" designation will stigmatize the use of coal ash in construction materials such as cement mix and wallboard, ultimately reducing its beneficial use and increasing the amount that must be disposed of.
RCRA Sub-title D-Solid Waste Management Requirements	CCW would be regulated as solid waste. Essentially, the waste would continue to be regulated pursuant to current requirements. That is, its disposal and use would be regulated in accordance with terms dictated by individual states. EPA could chose to develop landfill criteria or permitting requirements under subtitle D, but those requirements would not be enforceable.	Many states and industry organizations favor this approach. States prefer to regulate the waste as they see fit. Industry groups argue that current solid waste requirements are sufficient	Environmental groups argue that sufficient protections to human health and the environment must involve restrictions on land disposal and include enforceable requirements. This approach would leave inspection and permitting to states, which could result in inconsistent approaches and make it difficult for EPA to enforce the rules (if EPA chooses to develop disposal criteria).
RCRA Section 7003- Imminent hazard protection	EPA would use its current authority to address potential hazards at individual disposal sites.	Industry groups argue that this approach would allow EPA to address disposal units at power plants as well as uses that were not truly "beneficial" (e.g., land application as fill material as opposed to its use as a component in concrete)	Environmental groups argue that this is a resource intensive approach to addressing the issue. It is not a regulatory approach intended to regulate disposal facilities nation wide, but instead a provision that allows EPA to act in the case of an emergency. CCW should be regulated in a way that prevents an emergency from happening.

Source: Table prepared by CRS based on provisions of RCRA and public statements from various interest groups regarding potential CCW regulation.

Note: There has also been discussion of a potential "hybrid approach" that would involve regulating the waste as a solid waste if it were disposed of under certain conditions (e.g., in a lined landfill), but as a hazardous waste if those conditions were not met. It is unclear if such an approach would be enforceable considering EPA's current authority under RCRA.

Considering the recent interpretation of its authority under subtitle D and the current existing authority (described above), it appears that EPA generally has three options for regulating CCW. Those options, as well as selected and pros and cons outlined by various stakeholders, are summarized in Table 3.

On December 17, 2009, EPA issued a statement indicating that the regulatory proposal is on hold.

REQUIREMENTS APPLICABLE TO CCW MANAGEMENT

As discussed previously, CCW is managed in one of two ways—it is either *disposed of* in landfills, surface impoundments, or mines, or it is put to some *beneficial use*. In 2008, 136 million tons of CCW were generated. Industry estimates indicate that 8% was disposed of in mines as minefill and 37% was used in some capacity (e.g., as a component in concrete, cement, or gypsum wallboard, or as structural or embankment fill).[29] The remainder was disposed of in landfills or surface impoundments. Requirements applicable to each of these management methods are determined by individual states. Generally, the only federal role in their management may be that certain state permit programs (e.g., those related to wastewater discharges to surface water) are implemented under the authority of federal law. Other than that, there is little federal role in CCW management.

It is difficult to make any broad statements about regulations applicable to CCW management methods. Regulations do not just vary from state to state, but from unit to unit. For example, a given state likely regulates surface impoundments and landfills under different requirements associated with regulations applicable to solid waste and wastewater management, while mine disposal and beneficial uses may be largely unregulated.

Regulatory requirements within a given state may also vary depending on when a disposal unit went into operation. For example, requirements applicable to older landfills may be grandfathered in, under less stringent requirements than newer units, as new laws are enacted. Also, no industry or federal agency tracks the total number of disposal units or waste usage sites (e.g., locations where waste may have been used as structural fill). That is not to say that the waste is necessarily unregulated or disposed of improperly—only that there are many unknown elements of both current disposal and use practices, and probably even less that is known about sites that have been closed.

Landfill and Surface Impoundment Disposal

Landfilling CCW involves the long-term disposal of generally dry waste that is placed on an area of land or an excavation for permanent disposal. Surface impoundment units hold liquid waste that is generally sluiced directly from a power plant to the impoundment unit, where solids settle out, leaving relatively clear water at the surface (which may be recirculated into the plant or discharged to surface water). The impoundment itself may be a natural or man-made depression or diked area formed of earthen materials used for temporary or permanent storage or treatment of liquid waste. Solids may accumulate until the impoundment unit is full, or they may be dredged periodically and taken to another disposal unit such as a landfill.

In 2006, in a joint-agency effort, EPA and the U.S. Department of Energy (DOE) conducted a study to determine state regulatory requirements applicable to CCW landfills and surface impoundments built between 1994 and 2004.[30] At the time, it was estimated that roughly two- thirds of the waste was disposed of in landfills and the remainder in surface impoundments. However, it is unknown how accurate that estimate is. In March 2009, when EPA surveyed power plants to determine the integrity of existing surface impoundments, the agency estimated that there were approximately 300 ponds nationwide. Instead, survey results found that there are 584. There is no comparable data on the number of landfills.

The EPA/DOE study also found that regulations varied for each type of surface disposal unit, and varied significantly from state to state. A common regulatory element was that all disposal units were required to have some type of permit to operate—generally more than one. The most commonly issued state permits were issued in accordance with a state's solid waste requirements, wastewater or water pollution control requirements, or dam safety requirements.

As noted in the EPA/DOE study, these requirements applied to "new" landfills and surface impoundments. It is unknown how many landfills and surface impoundments built before 1994 exist, are still in operation, or may not have been properly closed. Older units may be required to have permits to continue operation, but would not likely have been required to install liners, leachate collection systems, or groundwater monitoring devices.

With regard to surface impoundments, states commonly regulate two elements of a unit—the structure itself (commonly pursuant to the state's dam safety requirements) and any discharges from the unit to surface or groundwater (commonly pursuant to the state wastewater or water pollution control requirements).

Another complicating factor in determining state CCW disposal requirements is states' tendency to allow a certain number of "exceptions" to state regulatory requirements. For example, a state may have specific requirements for landfill and surface impoundment liners or groundwater monitoring systems, but allow an individual plant to implement an alternative means of compliance on a case-by-case basis.

With regard to both surface impoundments and landfills, many states are likely to require groundwater monitoring to detect contamination from a disposal unit, but fewer states are likely to have regulatory requirements intended to *prevent* groundwater contamination from occurring (e.g., they would not likely require a plant to install a liner in an older, unlined landfill). These potential variations within a state's own program make comparison from state to state even more difficult. This variation, in part, was the basis for EPA's 2000 determination that consistent, national regulation regarding CCW disposal under RCRA Subtitle D was needed.

Although details of each state's regulatory requirements vary, there are certain broad requirements that may be similar. For example, a state is likely to regulate new landfills under provisions of the state's solid waste management program. New surface impoundments are likely regulated under provisions of the state's dam safety program and under the terms of a wastewater discharge permit program.

Solid Waste Permits

States may regulate CCW landfills in accordance with state solid waste management program requirements. Most state waste management programs specifically exclude CCW from the definition of hazardous waste and instead regulate it as solid waste. States generally regulate solid waste disposal in accordance with a permit program. That is, landfills generally are required to operate in accordance with criteria specified in a permit. If a state regulates CCW landfills under its solid waste permit program, new CCW landfills likely are required to have a liner and groundwater monitoring system. Permits may also require leachate collection systems, closure and post-closure requirements, siting controls, or a financial assurance requirement.

Not all states regulate CCW landfills through a permit program. For example, in the EPA/DOE study, of the 11 state programs analyzed, five adopted laws and regulations that resulted in exemptions from solid waste permitting requirements for certain CCW landfills. That does not necessarily mean that those states exempt CCW landfills from regulation, just that operational requirements established by the state are not met by complying with the terms of a permit.

Dam Safety Requirements

Many states use their dam safety requirements to regulate the construction, operation, and maintenance of surface impoundments. Such requirements would be intended to prevent a breach of a unit (such as the breach that occurred at Kingston). State dam requirements applicable to CCW surface impoundments may be similar to those for *mining waste* surface impoundment requirements found in the Surface Mining Control and Reclamation Act (SMCRA)—specifically requirements applicable to the unit design, construction, inspections, and emergency reporting. For example, West Virginia, a state that generates a significant amount of CCW, has dam safety requirements applicable to CCW surface impoundments. Selected elements of those requirements are:

- Units must have an application on file and a certificate of approval from the state to place, construct, or perform major repairs of a waste disposal dam.
- Plans and specifications of the design and construction must include, among other information, data regarding existing site conditions, subsidence potential, routine inspection and maintenance procedures and schedules, sediment control measures, the placement of spillways, seeding and mulching of the project area, surface drainage structures, installation of reading and monitoring devices, and an inventory of protected sites.[31]
- Units must meet specific design requirements, such as conformance to general hydrological requirements and, like SMCRA, address criteria applicable to foundation stability, structural consideration, and spillways.
- Units must meet construction requirements regarding inspections, operations and safety (including emergency procedures), and maintenance.

The presence of strong dam safety requirements is not a guarantee that regulated units will actually *be* operated and maintained according to those requirements. The requirements may be only as strong as a state's ability to enforce them. A state may have hundreds of structures required to meet its dam safety requirements and only a limited number of inspectors to insure that they are operated or maintained in an appropriate manner. In addition, a state may have strict requirements applicable to dam construction and operation, but limited ability to inspect those dams as often as necessary to ensure compliance. This makes it almost impossible to gauge the degree to which states are able to enforce their requirements.

Wastewater Discharge Permits

Disposal units, particularly surface impoundments, may be regulated as water pollution control facilities (as opposed to solid waste management units, such as landfills). In general, water pollution control facilities treat or store wastewater, including industrial wastewater, and discharge it directly or indirectly into the waters of a state, which may encompass both surface water and groundwater located wholly or partly within the state.

A disposal unit that has an outfall that discharges to surface water would be required to meet effluent guidelines specified under requirements of the Clean Water Act (CWA), and to operate in accordance with parameters specified in a National Pollutant Discharge Elimination System (NPDES) permit. As a federal requirement, all states that have been authorized by EPA to administer the NPDES program are required to regulate discharges to surface waters in accordance with certain minimum requirements (e.g., in accordance with federally mandated effluent standards). Specifics regarding how a permit program is implemented may vary (as long as minimum federal requirements are adhered to). For example, state water quality agencies may evaluate facilities on a case-by-case basis to determine the need for groundwater-protection measures such as impoundment liners and groundwater monitoring.

Even facilities that do not discharge wastewater to surface or groundwater may still be regulated in accordance with alternative water pollution control permits. Such facilities may be evaluated on a case-specific basis to determine the need for groundwater protection measures such as liners and groundwater monitoring.

Mine Disposal

The Department of the Interior's Office of Surface Mining (OSM) administers provisions of the Surface Mining Control and Reclamation Act (SMCRA). Among other provisions, SMCRA specifies requirements applicable to mine reclamation. CCW can be used in the reclamation process when it is used as minefill. Potential benefits associated with the use of CCW include its potential to abate acid mine drainage (due to the alkalinity of much of the waste), to improve already-disturbed mine lands, and to avoid increased generation of aboveground landfills and surface impoundments.[32]

As with the disposal of CCW in landfills and surface impoundments, there are no explicit federal requirements specific to the use of CCW as part of the mine reclamation process. However, unlike landfill and surface impoundment disposal, some states define minefill disposal as a beneficial use that is exempt from any regulation or restriction.

In its May 2000 regulatory determination, EPA stated that regulations under Subtitle D of RCRA (and/or possibly modifications to existing regulations established under the authority of the SMCRA) were warranted when these wastes are used to fill surface or underground mines.[33] In 2000, minefill disposal was a relatively new practice that lacked long-term monitoring data regarding its potential risks. In 2003, Congress requested that EPA commission an independent study of the health, safety, and environmental risks associated with the placement of CCW in active and abandoned coal mines in all major U.S. coal basins. As a result, the National Research Council (NRC) established the Committee on Mine Placement of Coal Combustion Wastes in September 2004.

In March 2006, the NRC committee published its study *Managing Coal Combustion Residues in Mines*. In part, it found that placing CCW in coal mines as part of the reclamation process is a viable management option as long as the waste placement is properly planned and carried out in a manner that avoids significant adverse environmental and health impacts, and that the regulatory process for issuing permits includes clear provisions for public involvement.[34]

The NRC committee cautioned that an integrated process of waste characterization, site characterization, management and engineering design of placement activities, and design and implementation of monitoring is required to reduce the risk of contamination moving from a mine site to the ambient environment. It stated further that comparatively little is known about the potential for minefilling to degrade the quality of groundwater and/or surface waters, particularly over longer time periods.

Table 4. Primary Beneficial Uses of Coal Combustion Wastes Encapsulated and Unencapsulated Uses

	CCW Product Description	Type of CCW Used
Encapsulated Uses		
Concrete	Concrete consists of a mixture of approximately 25% fine aggregate (sand), 45% gravel, 15% Portland cement, and 15% water.	Certain types of fly ash can replace a percentage of the Portland cement component of concrete, and are typically less expensive than Portland cement.
Cement additive	Cement clinker is an intermediary product of the Portland cement manufacturing process. Clinker is formed when a raw mix consisting of limestone, clay, bauxite, iron ore and quartz are heated in a kiln at higher temperatures.	Fly ash can be blended with limestone or shale and fed into the cement kiln to make clinker, which is then ground into Portland cement. FGD gypsum can be used to offset virgin gypsum in cement manufacture.
Gypsum wallboard	Gypsum wallboard (or drywall) is used as an interior finish in the construction of homes and building. Wallboard is composed of a layer of gypsum stucco sandwiched between two sheets of heavy paper.	FGD gypsum can replace 100% of virgin gypsum in wallboard after the excess moisture has been removed.
Road base	A road base is a foundation layer underlying a pavement and overlaying a subgrade of natural soil or embankment fill material. It protects the underlying soil from the detrimental effects of weather conditions and from the stressses and strains induced by traffic loads.	Bottom ash can be used to offset virgin sand or gravel in road base.

Table 4. (Continued)

	CCW Product Description	Type of CCW Used
Unencapsulated Uses		
Structural fill/ embankments	Structural fill is an engineered material used to raise or change the surface contour of an area and to provide ground support beneath building foundations. It can also be used to form embankments.	Depending on the soil type, fly ash can replace a percentage (generally 50%) of virgin rock, dirt, sand, or gravel in structural fill. Bottom ash can be used to offset virgin sand and gravel in structural fill.
Waste stabilization	CCW can be used in place of Portland cement, cement kiln dust, or lime to solidify and harden wet or liquid waste before it is landfilled.	Certain types of fly ash harden by themselves in contact with moisture; others can be mixed with another hardening agent, such as Portland cement, in order to be used in waste stabilization.
Soil modification/ stabilization/a gricultural uses	Gypsum (calcium sulfate dihydrate) can be used as a nutrient source for crops; as a conditioner to improve soil's physical properties and water infiltration and storage; to remediate sodic (high-sodium) soils; and to reduce nutrient and sediment movement to surface waters.	Fly ash and flue gas desulfurization gypsum (a synthetic material of identical chemical structure as natural, mined gypsum) can be used as a soil amendment to neutralize acidic soils.

Source: Table prepared by CRS based on data from EPA's Office of Solid Waste report, "Waste and Materials- Flow Benchmark Sector Report: Beneficial Use of Secondary Materials-Coal Combustion Products," February 1 2, 2008, pp. 2-5 through 2-6; EPA brochure, "Agricultural Uses for Flue Gas Desulfurization (FGD) Gypsum," EPA530-F-08-009; and ACAA, "2007 Coal Combustion Product (CCP) Production & Use Survey Results."

The committee recommended the establishment of enforceable federal standards to govern the placement of CCW in mines. The committee's reasoning for its recommendation, after reviewing the laws and other relevant literature, was that, although SMCRA does not specifically regulate CCW placement at mine sites, its scope is broad enough to encompass such regulation during reclamation activities. Further, while SMCRA and its implementing regulations indirectly establish performance standards that could be used to regulate the manner in which CCW may be placed in coal mines, neither the statute nor those rules explicitly addresses regulation of the use or placement of CCW, and some states have expressed concern that they do not have the authority to impose performance standards specific to CCW. Therefore, the committee recommended that enforceable federal standards be established for disposal of CCW in mines. It proposed that OSM regulations be changed to address CCW specifically, or that joint rules be developed by OSM and EPA under the authority of both SMCRA and RCRA.

"Beneficial Use"

In 2008, according to industry, approximately 37% of CCW was used in some capacity—most commonly as a component in concrete products, blended cement, gypsum panel products, and structural fill.[35] Some types of CCW are also used for road-base materials, roofing tiles and shingles, snow and ice control, and soil modification.

In its 1993 and 2000 regulatory determinations, among other factors, EPA looked at:

- Alternatives to current disposal methods.
- The costs of such alternatives.
- The impact of those alternatives on the use of natural resources.
- The current and potential utilization of coal combustion products.

After its analyses, EPA did not identify any environmental harm associated with the beneficial use of coal combustion products, and concluded in each regulatory determination that these materials did not warrant regulation as hazardous waste.

The beneficial use of coal combustion products can include both encapsulated and unencapsulated applications. The potential for contaminants to leach from CCW products largely depends on whether the waste is bound or encapsulated—as it would be in construction materials. According to EPA, unencapsulated uses of CCW require proper hydrogeologic evaluation to ensure adequate groundwater protection.[36] **Table 4** describes the primary encapsulated and unencapsulated uses of CCW.

Under EPA's Resource Conservation Challenge (RCC), CCW is an industrial material targeted for increased use as a building and manufacturing material. As part of that effort, EPA formed the Coal Combustion Products Partnership (C2P2) program with the American Coal Ash Association, Utility Solid Waste Activities Group, DOE, U.S. Department of Agriculture's Agricultural Research Service, Department of Transportation's Federal Highway Administration, and Electric Power Research Institute to help promote the beneficial use of CCW and "the environmental benefits that result from their use."[37]

EPA has been criticized for promoting certain "beneficial uses" of CCW without first determining if such uses are safe. In particular, the safety of agricultural uses or its use as structural or embankment fill has been questioned. National attention[38] was brought to its use as a fill material after developers used at 1.5 million tons of dry fly ash to build a golf course over a shallow aquifer at the Battlefield Golf Course, in Chesapeake, Virginia.

An assessment of groundwater wells and private drinking water wells in close proximity to the golf course found elevated levels of arsenic, barium, chromium, copper, iron, lead, mercury, and zinc.[39] However, from the available data, it has not been determined conclusively that the fly ash placed on the site has impacted nearby residential wells. Monitoring of the site is ongoing.

Issues associated with the Battlefield Gold Course site brought attention to what little is known about certain "beneficial" uses of CCW. In November 2009, the EPA Office of Inspector General issued its findings in an investigation into allegations of a cover-up in the risk assessment for the coal ash rulemaking. The Inspector General found no evidence of wrongdoing, but in that report stated that it has opened an investigation into the EPA's "partnership" with the coal industry to market coal ash and other combustion wastes in consumer, agricultural, and industrial products.[40] The report recommended a new probe of why EPA was promoting coal ash prior to determining whether these commercial applications were prudent or safe. In part, the report stated:

We identified a potential issue related to EPA's promotion of beneficial use through its Coal Combustion Product Partnership and have referred the question how EPA established a reasonable determination for these endorsements to the appropriate OIG office for evaluation.

CONCLUSION

Since the regulation of CCW disposal and use is controlled by individual states, it is difficult to determine certain information about the waste. For example, it is difficult to determine the entire amount of CCW that has been disposed of in the United States. It can be estimated (although not known definitively) how many *currently operational* disposal units exist today, but is not likely possible to determine the total number that have *ever been* in operation—that is, unlined units that may have been closed without a cap or groundwater monitoring system. Also, it is difficult to determine the number of sites that have used unencapsulated CCW that have properly evaluated the site, as recommended by EPA, to ensure adequate groundwater protection.

As power plant emission standards become more stringent, and air emission control devices capture more contaminants, both the total waste generated and the total amount of toxins in them can be expected to increase. As regulations are formulated to address new or expanded landfills, there are still many questions unanswered regarding the controls in place to minimize the potential risks posed by existing facilities—both with regard to a sudden, catastrophic release (as that in Kingston) or a gradual release and migration of contaminants.

Congressional interest in the issue existed before the Kingston release, but increased significantly afterward. On January 14, 2009, the Coal Ash Reclamation, Environment, and Safety Act of 2009 (H.R. 493) was introduced. The bill was intended to establish new standards applicable to surface impoundments. On February 12, 2009, the House Committee on Natural Resources, Subcommittee on Energy and Mineral Resources, held a legislative hearing on the bill. A scheduled markup was canceled when EPA announced that it would soon propose new regulations applicable to landfills and surface impoundments. The House Transportation and Infrastructure Committee, Subcommittee on Water Resources and Environment, held several hearings that looked at different aspects of the Kingston release, such as potential water quality impacts, causes of the release, and cleanup progress.[41] Also, on December 10, 2009, the House Committee on Energy and Commerce, Subcommittee on Energy and the Environment, held a hearing entitled "Drinking Water and Public Health Impacts of Coal Combustion Waste Disposal." Given its interest in this issue, it is unclear how Congress may respond given the current debate regarding EPA's existing authority to regulate CCW and its potential to regulate it as hazardous waste.

End Notes

[1] In this chapter, waste *management* generally refers to any method of handling waste after it has been generated. With regard to CCW, it refers to the *disposal* (in landfills, surface impoundment, or mines) or *use* of the waste (e.g., as a component in cement or concrete).

[2] See "EPA's Response to the TVA Kingston Fossil Plant Fly Ash Release," regarding January 6, 2009, sampling activity at http://www.epa.gov/region4/kingston/index.html.

[3] TVA "Form 10-Q" financial report for the quarterly period ended June 30, 2009, filed with the U.S. Securities and Exchange Commission on July 31, 2009, p. 45, available at http://investor.shareholder.com/tva/sec.cfm.

[4] EPA's Office of Solid Waste, "Human and Ecological Risk Assessment of Coal Combustion Wastes," August 6, 2007.

[5] See the House Committee on Natural Resources, Subcommittee on Energy and Mineral Resources hearing, "How Should The Federal Government Address The Health And Environmental Risks Of Coal Combustion Waste?," June 10, 2008. Also, on February 12, 2009, the subcommittee held a legislative hearing regarding H.R. 493, the "Coal Ash Reclamation, Environment, and Safety Act of 2009." On March 31, 2009, the House Committee on Transportation and Infrastructure, Subcommittee on Water Resources and Environment, held a hearing, "The Tennessee Valley Authority's Kingston Ash Slide: Potential Water Quality Impacts of Coal Combustion Waste Storage."

[6] Environmental Protection Agency, "Regulatory Determination on Wastes from the Combustion of Fossil Fuels; Final Rule," 65 *Federal Register* 322 14-32237.

[7] See EPA's "Statement From EPA On Coal Ash," issued December 17, 2009. EPA's press releases are available, by date, at http://www.epa.gov/newsroom/.

[8] The term "beneficial use" is not specifically defined by EPA (although it may be defined in individual state regulations). It is generally meant to include uses of CCW that would provide some environmental, economic, or performance benefit, when compared to direct disposal of the waste.

[9] The National Research Council (NRC), *Managing Coal Combustion Residues in Mines*, March 2006, pp. 27-57.

[10] See U.S. Geological Survey, Energy Resources Program, available at http://energy coal _databases.html.

[11] For more information, see the NRC report, *Managing Coal Combustion Residues in Mines*, pp. 59-76.

[12] EPA, Office of Solid Waste, "Coal Combustion Waste Damage Case Assessment," July 9, 2007.

[13] EPA requested this information pursuant to its authority under § 104(e) of the Comprehensive Environmental Response, Compensation, and Liability Act (CERCLA, 42 U.S.C. § 9604(e)), which provides that when the agency has reason to believe that there may be a release or threat of a release of a pollutant or contaminant, it may require any person who has or may have information about the release to furnish information relating to the matter to EPA. For more information about EPA's request for information, and the response from utilities, see EPA's "March 9, 2009 Coal Ash Information Request Letter," at http://www.epa.gov/osw/ nonhaz/industrial/special/fossil/coalashletter.htm.

[14] For more information, see EPA's "Fact Sheet: Coal Combustion Residues (CCR) - Surface Impoundments with High Hazard Potential Ratings," document EPA530-F-09-006, available at http://www.epa.gov/osw /nonhaz/industrial/ special/fossil/ccrs-fs/index.htm.

[15] RCRA actually amends earlier legislation, the Solid Waste Disposal Act of 1965, but the amendments were so comprehensive that the act is commonly referred to as RCRA rather than by its official title.

[16] Solid waste is defined in more detail at 40 C.F.R. 261.2.

[17] Hazardous waste is a subset of solid waste. A waste must first be determined to be a solid waste before it can meet the definition of hazardous waste.

[18] Other test methods may also be acceptable. For more information, see EPA's website regarding various test methods for evaluating solid waste, at http://www.epa.gov/waste

[19] See 40 C.F.R. §§ 261.31, 261.32, and 261.33.

[20] For more information about the identification of solid and hazardous waste and hazardous waste exclusions, see the Colorado Department of Public Health and Environment, Hazardous Materials and Waste Management Division, "Hazardous Waste Exclusions Guidance," Second Edition, April 2009, available at http://www.cdphe.state.co.us/HM/ hwexcl.pdf.

[21] 42 *Federal Register* 58946, December 18, 1978.

[22] For details on EPA's Office of Enforcement and Compliance Assurance, see "Guidance on the Use of Section 7003 of RCRA," October 1997, available at http://www.p2pays.org/ref/03/02645.pdf. For information on legal requirements for initiating action under Section 7003, in particular, see pp. 9-19.

[23] *Congressional Record*, February 20, 1980, p. 1087.

[24] 45 *Federal Register* 33084, May 19, 1980.

[25] Section 8002(f), (n) and (p) of RCRA.

[26] Ibid.

[27] 42 U.S.C.§ 6907(a), "Solid Waste Management Information and Guidelines" and §6944(a), "Upgrading of Open Dumps."

[28] *InsideEPA*, "Environmental Policy Alert: EPA Lawyers Stymie State, Industry Bid For 'Solid' Waste Coal Ash Rules," October 17, 2009.

[29] See the American Coal Ash Association (ACAA), "2008 Coal Combustion Product (CCP) Production & Use Survey Results (Revised)," available at http://www.acaa-usa.org/displaycommon.cfm?an=1&subarticlenbr=3. The ACAA considers "mining applications" a "use" of CCW. The extent to which such applications are actually minefill is not defined. In this chapter, the use of CCW as minefill is considered another method of disposal and therefore is not included in statistics regarding "beneficial use."

[30] EPA and the U.S. Department of Energy report, *Coal Combustion Waste Management at Landfills and Surface Impoundments, 1994-2004*, August 2006.

[31] A complete list of plan requirements is specified under the West Virginia state regulations at 47 C.S.R., 34 §6.

[32] Truett Degeare, U.S. Environmental Protection Agency, Office of Solid Waste, presentation, "Overview of U.S. Environmental Protection Agency Coal Combustion Waste (CCW) Mine Fill Issues," available at http://aciddrainage.com/ps/ccb2/5-1.pdf.

[33] EPA's data collection and analysis efforts associated with developing regulations applicable to CCW mine disposal have proceeded on a track separate from the efforts to develop regulations associated with CCW disposal in landfills and surface impoundments.

[34] Sections of the NRC report referenced here are taken largely from the Department of the Interior's Office of Surface Mining Reclamation and Enforcement's March 14, 2007, Advance Notice of Proposed Rulemaking, "Placement of Coal Combustion Byproducts in Active and Abandoned Coal Mines," 72 *Federal Register* 12025-12030.

[35] The ACAA, "2007 Coal Combustion Product (CCP) Production & Use Survey Results." In determining reuse totals, ACAA considers "mining applications" a reuse of CCW. The extent to which such applications are actually minefill is not defined. In this CRS report, the use of CCW as minefill is considered another method of disposal and therefore not included in statistics regarding reuse.

[36] See EPA's Coal Combustion Products-Regulatory Resources webpage: http://www.epa.gov/wastes /conserve/rrr/imr/ ccps/resources.htm.

[37] For more information, see the C2P2 website at http://www.epa.gov/epawaste/partnerships/c2p2/index.htm.

[38] See *CBS News, 60 Minutes,* "Coal Ash: 130 Million Tons of Waste," originally aired October 4, 2009, available at http://www.cbsnews.com/stories/2009/10/01/60minutes/main5356202.shtml.

[39] See "Draft Site Inspection for Battlefield Golf Club Site, Chesapeake, Virginia," prepared for EPA Region 3 by Tetra Tech EM, Inc., March 30, 2009, p. 10, available at http://www.chesapeake.va.us/SERVICES/ citizen_info/ battlefieldgolfclub/pdf/Draft-Battlefield-Golf-Club-SI.pdf.

[40] EPA Office of the Inspector General, "Response to EPA Administrator's Request for Investigation into Allegations of a Cover-up in the Risk Assessment for the Coal Ash Rulemaking," Report No. 10-N-0019, November 2, 2009, available at http://www.epa.gov/oig/reports/2010/20091102-10-N-0019.pdf.

[41] "The Tennessee Valley Authority's Kingston Ash Slide: Potential Water Quality Impacts of Coal Combustion Waste Storage," on March 31, 2009, "Coal Combustion Waste Storage and Water Quality, " on April 30, 2009, "The Tennessee Valley Authority's Kingston Ash Slide: Evaluation of Potential Causes and Updates on Cleanup Efforts," July 28, 2009, and "The One Year Anniversary of the Tennessee Valley Authority's Kingston Ash Slide: Evaluating Current Cleanup Progress and Assessing Future Environmental Goals," on December 9, 2009.

In: Coal Combustion Waste: Management and Beneficial Uses ISBN: 978-1-61728-962-0
Editor: Daniel D. Lowell

Chapter 2

WASTE AND MATERIALS-FLOW BENCHMARK SECTOR REPORT: BENEFICIAL USE OF SECONDARY MATERIALS-COAL COMBUSTION PRODUCTS

United States Environmental Protection Agency

ACKNOWLEDGMENTS

Industrial Economics, Incorporated (IEc), is responsible for the overall organization and development of this chapter. This chapter was developed with the assistance of Dr. H. Scott Mathews of the Carnegie Mellon Green Design Institute, Dr. Jim Boyd of Resources for the Future, Dave Goss of the American Coal Ash Association (ACAA), and with input and data from various EPA Office of Solid Waste and Office of Policy Economics and Innovation representatives. Lyn D. Luben of the U.S. Environmental Protection Agency, Office of Solid Waste, provided guidance and review.

EXECUTIVE SUMMARY

Introduction

The U.S. Environmental Protection Agency's Office of Solid Waste (EPA OSW) is currently developing methods to evaluate the environmental, human health, and economic outcomes of specific EPA programs. As an initial step, OSW is examining the extent to which the costs and benefits of source reduction, reuse, and recycling may be quantified for a range of materials targeted by the Resource Conservation Challenge (RCC).

Coal combustion products (CCPs) are among the materials targeted by EPA's Resource Conservation Challenge (RCC). The RCC is designed to facilitate changes in the economics and practice of waste generation, handling, and disposal (e.g., by promoting market opportunities for beneficial use). Under the RCC, EPA has established three goals for increased beneficial use of CCPs:

Exhibit ES-1. ACAA Survey of Key Beneficial Use Applications for CCPS in 2005 (Million Short Tons)

APPLICATION (INDUSTRY)	COAL FLY ASH	BOTTOM ASH	FGD GYPSUM	OTHER FGD WET MATERIAL	FGD DRY MATERIAL	BOILER SLAG	FBC ASH	TOTAL
Concrete[a] (Construction)	14.99	1.02	0.33	0	0.01	0	0	**16.35**
Structural fill[b] (Construction)	5.71	2.32	0	0	< 0.01	0.18	0.14	**8.35**
Wallboard[c] (Construction)	0	0	8.18	0	0	0	0	**8.18**
Raw feed for cement clinker[d] (Construction)	2.83	0.94	0.40	< 0.01	0	0.04	0	**4.22**
Waste stabilization[e] (Waste Mgmt)	2.66	0.04	0	0	0	0	0.14	**2.84**
Blasting Grit/Roofing Granules	0	0.89	0	0	0	1.54	0	**1.63**
Total - Key Uses	***26.19***	***4.41***	***8.90***	***< 0.01***	***0.02***	***1.76***	***0.28***	***41.57***
Total – Other Uses[f]	***2.93***	***3.13***	***0.36***	***0.69***	***0.014***	***0.13***	***0.66***	***8.04***
TOTAL - ALL USES	**29.12**	**7.54**	**9.27**	**0.69**	**0.16**	**1.89**	**0.94**	**49.61**
2005 QUANTITY GENERATED	**71.10**	**17.60**	**12.00**	**17.70**	**1.43**	**1.96**	**1.37**	**123.13[g]**
CCP UTILIZATION RATE[h]	**41%**	**43%**	**77%**	**4%**	**11%**	**97%**	**69%**	**40%**

Notes:

a. CCPs are frequently used as a replacement for a portion of portland cement in the manufacture of concrete.

b. Structural fill is an engineered material that is used to raise or change the surface contour of an area and to provide ground support beneath highway roadbeds, pavements and building foundations. It can also be used to form embankments.

c. FGD gypsum is used as a substitute for virgin gypsum in wallboard manufacturing.

d. CCPs can be blended with limestone or shale and fed into the cement kiln to make clinker, which is then ground into portland cement.

e. The chemical properties of CCPs make them effective stabilizers of biosolids (i.e., sludge from municipal waste water treatment).

f. Includes quantities beneficially used in minor applications not included in this exhibit, but listed in Appendix A.

g. Includes 115,596 tons of "Other FGD Material" not listed in this table because of the small quantities generated.

h. CCP utilization rates reflect all use applications, some of which are omitted from this table but are included in Appendix A. Utilization rates are calculated by dividing the total quantity used by the total quantity generated.

Note: Results from the 2006 CCP Production and Use Survey conducted by the ACAA indicate a total utilization rate of 43.43 percent, up from 40.29 percent reported for 2005. This reflects an ongoing upward trend in the CCP utilization rate over the past decade. The 2006 results were received too late for incorporation into this chapter.

Sources:

1. American Coal Ash Association. "2005 Coal Combustion Product (CCP) Production and Use Survey," accessed at: http://www.acaa-usa.org/PDF/20045_CCP_Production_ and_Use_Figures_ Released_ by_ACAA.pdf.
2. Western Region Ash Group, "Applications and Competing Materials, Coal Combustion Byproducts," accessed at: http://www.wrashg.org/compmat.htm.

- Achieve a 50 percent beneficial use rate of CCPs by 2011;
- Increase the use of coal fly ash in concrete by 50 percent (from 12.4 million tons per year in 2001 to 18.6 million tons by 2011); and
- Reduce greenhouse gas emissions from concrete production by approximately 5 million metric tons CO_2 equivalent by 2010.[1]

CCPs are formed during coal-burning processes in power plants and industrial boilers. Coal combustion produces various forms of CCPs, which are categorized by the process in which they are generated. Common CCPs include: fly ash, bottom ash, Flue Gas Desulphurization (FGD) material, boiler slag, Fluidized Bed Combustion (FBC) ash, and cenospheres. CCPs may be beneficially used as a component of building materials or as a replacement for other virgin materials such as sand, gravel, or gypsum. Size, shape, and chemical composition determine the suitability of these materials for beneficial use. Higher value applications, such as use in cement or concrete products, require moderately stringent specifications (in terms of size, shape and chemical composition), whereas lower value uses, such as structural or mining fills, can accept more variable materials.

This chapter serves two purposes: (1) To provide an initial assessment of the market dynamics that affect the generation, disposal, recovery, and beneficial use of CCPs; and (2) to provide a preliminary life cycle analysis of the beneficial impacts of CCP use, including an initial estimate of the baseline beneficial use impacts with current (2005) CCP levels and, for some materials, the beneficial impacts associated with achieving the 2011 RCC goal.

CCP Generation and Market Dynamics

The American Coal Ash Association (ACAA), a trade association whose purpose is to advance the beneficial use of CCPs, reports that the electric power industry generates approximately 123 million short tons of CCPs annually. Of these, the industry disposed of approximately 74 million short tons to landfills, while beneficially using approximately 50 million short tons in products.[2] Exhibit ES-1 summarizes results of the most recent (2005) ACAA survey of generators of CCPs, which indicates that the most common beneficial use applications for CCPs are as a replacement for virgin materials in concrete and cement-making, structural fill, and gypsum wallboard.

The CCP beneficial use market is composed of three primary segments. These are:

- **Generators:** Approximately 400 to 500 coal-fired electric utilities currently operate in the United States. Since the coal power industry consumes approximately 92% of all U.S. coal, it is responsible for producing the vast majority of CCPs in the country. Other industries that use coal as a fuel source in commercial or industrial boilers (e.g., mineral and grain processors) also produce small quantities of CCPs. Several factors influence a generator's decision to either dispose or seek beneficial use options for spent CCPs. Key considerations include the costs of landfill disposal, transport, processing, storage, and marketing.

- **Intermediaries:** Some coal-fired utilities market CCPs for beneficial use through a third-party instead of selling directly to users. In these cases, a utility perceives an efficiency in outsourcing the marketing of its CCPs. Marketers typically accept all of a generator's CCPs as a service to the company, sell the marketable portion, and dispose of the portion that is not salable. The marketer typically bears the cost of hauling CCPs from the utility and the liability associated with moving or storing the materials.
- **End-Users:** Several economic factors determine an end-user's decision to use CCPs in its product. These factors include: the price of CCPs relative to the price of virgin materials for specific uses; the technical fit between CCPs and the use application; access to sufficient quantities of CCPs; and federal and state policies associated with CCP use.

Impacts of Current Policy Setting on Market Dynamics

While states play a primary role in establishing industrial waste regulations and guidance, EPA has an opportunity to provide coordination and assistance at the national and regional level to help achieve a shift in waste management policy. EPA is currently engaged in several partnerships to facilitate and increase beneficial use of CCPs. Efforts within these partnerships include: promoting the beneficial use of CCPs through the development of web resources; developing technical guidance on the best practices for the beneficial use of CCPs; holding educational workshops and outreach support for CCP users; and providing recognition for the innovative beneficial use of CCPs. Key partners in these efforts include the American Coal Ash Association (ACAA), Utility Solid Waste Group (US WAG), the U.S. Department of Energy (DOE), and the Federal Highway Administration (FHWA).

Estimating the Impacts of the Beneficial Use of CCPs

To quantify the environmental impacts of increased beneficial use of CCPs in various applications, we use a life cycle analysis approach, as both a first step in an economic analysis, and, where economic analysis is not practical, as a meaningful proxy.

To estimate beneficial impacts of CCP use, we first develop preliminary estimates of the incremental impacts associated with using a specific quantity (e.g., one ton) of CCPs in different applications. These impacts can then be extrapolated in specific scenarios designed to address program-level outcomes. To fully capture the beneficial impacts of EPA program achievements, it is necessary to model each beneficial use application of all CCPs targeted by the RCC. However, the time, data, and resources required to perform this task are beyond the scope of this chapter. For this preliminary analysis, therefore, we have selected two common CCPs, fly ash and FGD gypsum, whose beneficial use applications are well understood, and for which life cycle models and existing data are available.

We conduct separate analyses to evaluate the incremental environmental impacts associated with beneficially using a specific quantity (e.g., one ton) of fly ash and FGD gypsum. We selected two life cycle modeling applications, Building for Environmental and Economic Sustainability (BEES) and SimaPro, to conduct the analyses. Both models have been peer-reviewed and evaluate a large suite of environmental metrics. We employ the

BEES model to investigate the beneficial impacts of using one ton of fly ash as a substitute for finished portland cement in concrete, and SimaPro to evaluate the use of one ton of FGD gypsum as a substitute for virgin gypsum in wallboard. Both analyses assume that the beneficial use material (fly ash or FGD gypsum) substitutes for virgin material (finished portland cement or virgin gypsum) on a one-to-one, mass-based basis. Exhibit ES-2 presents the results of the BEES and SimaPro analysis.

The results of the fly ash and FGD gypsum analyses suggest many positive environmental impacts from beneficial use. For most metrics, there is a significant difference between the unit impact value for fly ash and FGD gypsum. The difference in unit impact values reflects different avoided processes when fly ash is used to offset portland cement versus when FGD gypsum is used to offset virgin gypsum. For example, the primary driver of benefits when fly ash is used in concrete is avoided raw materials extraction and avoided portland cement production.[3] In comparison, the primary driver of benefits when FGD gypsum is used in wallboard is avoided virgin gypsum extraction and the processing of virgin gypsum into stucco. Portland cement production generates relatively high greenhouse gas emissions. Thus, the avoided CO_2 and methane emissions are greater for fly ash than for FGD gypsum in this analysis. In contrast, gypsum mining requires comparatively higher quantities of water, so the water savings are greater for FGD gypsum in this analysis than for portland cement. In addition, the difference in unit impacts likely reflects minor differences in the system boundaries in each analysis and the data sets utilized by each model.

Estimating Program Level Impacts

In order to extrapolate the beneficial impacts presented in Exhibit ES-2 to evaluate EPA's program level efforts, two critical steps are necessary.

Exhibit ES-2. Incremental Beneficial Impacts of Using Fly Ash in Portland Cement and FGD Gypsum in Wallboard

AVOIDED IMPACTS	PER 1 TON FLY ASH AS PORTLAND CEMENT SUBSTITUTE IN CONCRETE	PER 1 TON FGD GYPSUM IN WALLBOARD
ENERGY USE		
NONRENEWABLE ENERGY (MJ)[a]	4,214.18	12,568.97
RENEWABLE ENERGY (MJ)[b]	43.55	13.69
TOTAL PRIMARY ENERGY (MJ)	4,259.29	12,582.66
TOTAL PRIMARY ENERGY (US$)[c]	119.26	352.31
WATER USE		
TOTAL WATER USE (L)	341.56	14,214.60
TOTAL WATER USE (US$)[d]	0.22	9.01
GREENHOUSE GAS EMISSIONS		
CO_2 (G)	636,170.21	77,754.24
METHANE (G)	539.49	175.51

Exhibit ES-2. (Continued)

AVOIDED IMPACTS	PER 1 TON FLY ASH AS PORTLAND CEMENT SUBSTITUTE IN CONCRETE	PER 1 TON FGD GYPSUM IN WALLBOARD
AIR EMISSIONS		
CO (G)	593.45	39.06
NOX (G)	1,932.48	168.02
SOX (G)	1,518.21	139.14
PARTICULATES GREATER THAN PM_{10} (G)	0.00	1,194.25
PARTICULATES LESS THAN OR EQUAL TO PM_{10} (G)	0.01	520.93
PARTICULATES UNSPECIFIED (G)	1,745.25	17.11
MERCURY (G)	0.04	0.00
LEAD (G)	0.03	0.03
WATERBORNE WASTES		
SUSPENDED MATTER (G)	13.96	23.60
BIOLOGICAL OXYGEN DEMAND (G)	3.07	21.87
CHEMICAL OXYGEN DEMAND (G)	26.00	24.71
COPPER (G)	0.00	0.02
MERCURY (G)	0.00	0.00
LEAD (G)	0.00	0.01
SELENIUM (G)	0.00	0.00
NONHAZARDOUS WASTE (KG)[e]	0.00	3.12

Notes:

a. Nonrenewable energy refers to energy derived from fossil fuels such as coal, natural gas and oil.

b. Renewable energy refers to energy derived from renewable sources, but BEES does not specify what sources these include.

c. In addition to reporting energy impacts in megajoules (MJ), we monetize impacts by multiplying model outputs in MJ by the average cost of electricity in 2006 ($0.0275/MJ), converted to 2007 dollars ($0.0280/MJ). The 2006 cost of energy is taken from the Federal Register, February 27, 2006, accessed at: http://www.npga.org/14a/pages/index.cfm?pageid=914. The cost was converted to 2007 dollars using NASA's Gross Domestic Product Deflator Inflation Calculator, accessed at: http://cost.jsc.nasa.gov/inflateGDP.html.

d. In addition to reporting water impacts in gallons, we monetize impacts by converting model outputs from liters to gallons and multiplying by the average cost per gallon of water between July 2004 and July 2005 ($0.0023/gal), converted to 2007 dollars ($0.0024/gal). The 2005 cost of water is taken from NUS Consulting Group, accessed at: https://www.energyvortex.com/files/NUS_quick_click.pdf. The cost was converted to 2007 dollars using NASA's Gross Domestic Product Deflator Inflation Calculator, accessed at: http://cost.jsc.nasa.gov/inflateGDP.html.

e. BEES reports waste as "end of life waste." In contrast, SimaPro reports "solid waste." It is not clear if these waste metrics are directly comparable as SimaPro does not specify whether "solid waste" refers to manufacturing waste, end-of-life waste, or both.

Exhibit ES-3. Extrapolated Impacts of the Beneficial Use of CCPs

AVOIDED IMPACTS	FLY ASH IN CONCRETE EXTRAPOLATED TO RCC GOAL (18.6 MILLION TONS)[a]	FLY ASH IN CONCRETE EXTRAPOLATED TO CURRENT USE (15.0 MILLION TONS)[b]	FGD GYPSUM IN WALLBAORD EXTAPOLATED TO CURRENT USE (8.2 MILLION TONS)[c]	PARTIAL SUM OF CURRENT USE BENEFICIAL IMPACTS[d]
ENERGY USE				
NONRENEWABLE ENERGY (MJ)[e]	78.4 billion	63.2 billion	102.8 billion	166.0 billion
RENEWABLE ENERGY (MJ)[f]	810.0 million	652.8 million	111.9 million	764.7 million
TOTAL PRIMARY ENERGY (MJ)	79.2 billion	63.8 billion	102.9 billion	166.7 billion
TOTAL PRIMARY ENERGY (US$)[g]	$2.2 billion	$1.8 billion	$2.9 billion	$4.7 billion
WATER USE				
TOTAL WATER USE (LITERS)	6.3 billion	5.2 billion	116.2 billion	121.4 billion
TOTAL WATER USE (US$)[h]	$4.0 million	$3.2 million	$73.7 million	$77.9 million
GREENHOUSE GAS EMISSIONS				
CO_2 (G)	11.8 trillion	9.5 trillion	0.6 trillion	10.2 trillion
METHANE (G)	10.0 billion	8.1 billion	1.4 billion	9.5 billion
TONS CO_2 EQUIVALANT[i]	13.2 million	10.6 million	0.7 million	11.5 million
AIR EMISSIONS				
CO (G)	11.0 billion	8.9 billion	0.3 billion	9.2 billion
NOx (G)	35.9 billion	29.0 billion	1.4 billion	30.3 billion
SOx (G)	28.2 billion	22.8 billion	1.1 billion	23.9 billion
PARTICULATES GREATER THAN PM_{10} (G)	0	0	9.7 billion	9.7 billion
PARTICULATES LESS THAN OR EQUAL TO PM_{10} (G)	0.2 million	.02 million	4.3 million	4.3 million
PARTICULATES UNSPECIFIED (G)	32.5 billion	26.1 billion	0.1 billion	26.3 billion
MERCURY (G)	714,000	576,000	8,000	584,000
LEAD (G)	523,000	421,000	235,000	656,000
WATERBORNE WASTES				
SUSPENDED MATTER (G)	259.6 million	209.2 million	193.0 million	402.2 million
BIOLOGICAL OXYGEN DEMAND (G)	57.1 million	46.1 million	178.8 million	1224.9 million
CHEMICAL OXYGEN DEMAND (G)	483.6 million	389.7 million	202.1 million	591.8 million
COPPER (G)	0	0	194,000	194,000
MERCURY (G)	1	0	3,000	3,000

Exhibit ES-3. (Continued)

AVOIDED IMPACTS	FLY ASH IN CONCRETE EXTRAPOLATED TO RCC GOAL (18.6 MILLION TONS)[a]	FLY ASH IN CONCRETE EXTRAPOLATED TO CURRENT USE (15.0 MILLION TONS)[b]	FGD GYPSUM IN WALLBAORD EXTAPOLATED TO CURRENT USE (8.2 MILLION TONS)[c]	PARTIAL SUM OF CURRENT USE BENEFICIAL IMPACTS[d]
LEAD (G)	0	0	65,000	65,000
SELENIUM (G)	3	2	2,000	2,000
NON-HAZARDOUS WASTE (KG)[j]	**0**	**0**	**25.4 million**	**25.4 million**

Notes:

a. We extrapolate the incremental impacts (i.e., impacts associated with use of 1 ton fly ash) to estimate impacts of attaining the RCC goal for the use of fly ash in concrete (18.6 million tons by 2011). To extrapolate, we multiply each of the incremental impacts calculated by the BEES model by 18.6 million.

b. We extrapolate the incremental impacts (i.e., impacts associated with use of 1 ton fly ash) to estimate the impacts of current beneficial use of fly ash in concrete (15.0 million tons). The current quantity of fly ash that is beneficially used as a substitute for finished portland cement in concrete is reported by ACAA's 2005 CCP Survey. We multiply each of the incremental impacts calculated by BEES by 15.0 million tons to extrapolate these impacts to reflect current use.

c. We extrapolate the incremental impacts (i.e., impacts associated with use of 1 ton FGD gypsum) to estimate the impacts of current beneficial use of FGD gypsum in wallboard (8.2 million tons). The current quantity of FGD gypsum that is beneficially used as a substitute for finished portland cement in concrete is reported by ACAA's 2005 CCP Survey. We multiply each of the incremental impacts calculated by SimaPro by 8.2 million to extrapolate these impacts to reflect current use.

d. Calculated as the sum of the fly ash and FGD gypsum current use extrapolations.

e. Nonrenewable energy refers to energy derived from fossil fuels such as coal, natural gas and oil.

f. Renewable energy refers to energy derived from renewable sources, but BEES does not specify what sources these include.

g. In addition to reporting energy impacts in megajoules (MJ), we monetize impacts by multiplying model outputs in MJ by the average cost of electricity in 2006 ($0.0275/MJ), converted to 2007 dollars ($0.0280/MJ). The 2006 cost of energy is taken from the Federal Register, February 27, 2006, accessed at: http://www.npga.org/14a/pages/index.cfm?pageid=914. The cost was converted to 2007 dollars using NASA's Gross Domestic Product Deflator Inflation Calculator, accessed at: http://cost.jsc.nasa.gov /inflateGDP.html.

h. In addition to reporting water impacts in gallons, we monetize impacts by converting model outputs from liters to gallons and multiplying by the average cost per gallon of water between July 2004 and July 2005 ($0.0023/gal), converted to 2007 dollars ($0.0024/gal). The 2005 cost of water is taken from NUS Consulting Group, accessed at: https://www.energyvortex.com/files/NUS_quick_click.pdf. The cost was converted to 2007 dollars using NASA's Gross Domestic Product Deflator Inflation Calculator, accessed at: http://cost.jsc.nasa.gov/inflateGDP.html.

i. Greenhouse gas emissions have been converted to tons of CO_2 equivalent using U.S. Climate Technology Cooperation Gateway's Greenhouse Gas Equivalencies Calculator accessed at: http://www.usctc gateway.net/tool/. This calculation only includes CO_2 and methane.

j. BEES reports waste as "end of life waste." In contrast, SimaPro reports "solid waste." It is not clear if these waste metrics are directly comparable as SimaPro does not specify whether "solid waste" refers to manufacturing waste, end-of-life waste, or both.

- Development of defensible beneficial use scenarios that accurately identify the extent to which different beneficial uses are likely to increase; and
- Implementation of a well-supported attribution protocol for assigning beneficial use impacts to specific EPA programs.

At this time, the data necessary to develop accurate beneficial use scenarios and to support a clear attribution of impacts are not sufficient to inform a detailed program analysis. In the absence of such data, we present a preliminary analysis of the total impacts associated with current (baseline) beneficial use patterns. While these impacts do not strictly reflect RCC program achievements, they represent the best available information on the environmental benefits of beneficially using certain CCPs, and reflect the impacts of all EPA, state, and industry efforts to increase CCP use to its 2005 level. The beneficial use impacts of current fly ash and FGD gypsum use are calculated by extrapolating the impacts identified in Exhibit ES-3 to the current quantity of each material beneficially used in each application. For fly ash, we also extrapolate the beneficial impacts associated with achieving the 2011 RCC goal—a 50% increase in fly ash use in concrete. Exhibit ES-3 presents the key impacts of the beneficial use of CCPs extrapolated to current use quantities. Note that the impacts presented in Exhibit ES-3 represent only a partial estimate of the total impacts of beneficially using CCPs. Beneficial use of fly ash as a substitute for finished portland cement in concrete and FGD gypsum in wallboard accounts for only 47% (23.2 million tons) of all beneficially used CCPs in 2005.

The results show that current beneficial use of fly ash in concrete and FGD gypsum in wallboard results in positive environmental impacts. The most significant impacts include energy savings and water use reductions. Energy savings associated with the use of fly ash and FGD gypsum totals approximately 167 billion megajoules of energy (or approximately $4.7 billion in 2007 energy prices). Based on the average monthly consumption of residential electricity customers, this is enough energy to power over 4 million homes for an entire year. Avoided water use totals approximately 121 billion liters or approximately $76.9 million in 2007 water prices).[4] This is roughly equivalent to the annual water consumption of 61,000 Americans.[5] The extrapolated beneficial impacts also include key impacts such as avoided greenhouse gas (11.5 million tons of avoided CO2 equivalent), and avoided air emissions (30.3 million kilograms of avoided NOx, and 23.9 million kilograms of SOx).

This chapter also presents a distributional screening analysis using the EIO-LCA model that indicates significant avoided environmental impacts from reductions in the demand for cement or virgin gypsum that are distributed across several economic sectors. From the perspective of energy and air emissions, cement manufacturing leads to large impacts, and is in general the largest source of emissions across the supply chain. Reducing the amount of cement produced by beneficially reusing products can lead to large supply chain-wide reductions of emissions. Comparatively, the impact of the substitution of FGD gypsum for virgin gypsum in wallboard manufacturing is less clear, as the model was not able to adequately represent the wallboard sector.[6]

The preliminary results of this initial analysis suggest that a more detailed evaluation of the beneficial impacts of the beneficial use of CCPs could assist EPA in the more specific estimation of the achievements of the RCC program. A more detailed analysis would require:

- The development of realistic and effective beneficial use scenarios that incorporate more detailed descriptions of markets, beneficial uses, and policies. Realistic scenarios should reflect key market dynamics and limits such as distance to markets and virgin material prices, and be able to assess the impacts of these dynamics on the growth potential for specific beneficial uses.
- The development of a methodology to attribute beneficial use impacts to specific EPA/RCC efforts and programs. A phased approach may be employed that initially assumes all impacts result from EPA actions. This assumption could then be refined to reflect specific strategies, policies, and other efforts, and link these, where possible, to specific changes in beneficial use practices and markets.
- The expansion of the assessment to include additional CCPs and beneficial use applications. This analysis only examines the beneficial impacts of substituting using fly ash for finished portland cement in concrete and substituting FGD gypsum for virgin gypsum in wallboard manufacturing. The two processes represent less than 50% of the total beneficial use of CCPs. Additional high volume applications that EPA may wish to analyze include: the use of fly ash as a raw feed in cement clinker; the use of boiler slag as blasting grit; and the use of various CCPs in structural fill and waste stabilization. In addition, the Agency may investigate the beneficial impacts of lower volume applications to identify those that may have potentially high incremental impacts.

1. INTRODUCTION

The U.S. Environmental Protection Agency's Office of Solid Waste (EPA OSW) is currently considering the strategic direction of solid and hazardous waste policy. As part of this effort, OSW is developing methods to evaluate the environmental, human health, and economic outcomes of specific EPA programs to support strategic planning and program evaluation. Three important areas of focus in this transition are:

- Measurement of materials flow and life cycle impacts related to waste minimization and materials recovery and reuse, including an emphasis on "upstream" resource conservation beneficial impacts;
- Documentation of the impacts of voluntary programs, including the various efforts and materials targeted by EPA's Resource Conservation Challenge (RCC);[7] and
- Development of data and approaches that can support annual performance reporting under the Government Performance and Results Act (GPRA) and OMB's Performance Assessment Rating Tool (PART) evaluations.

As an initial step in the development of methods to assess the beneficial impacts of program benefits for both voluntary programs and PART, OSW is examining the extent to which the costs and benefits of source reduction, reuse, and recycling may be quantified for a range of materials targeted by the RCC.

This chapter examines one of the materials targeted under the RCC: coal combustion products (CCPs). CCPs are produced during coal-burning process at electric utilities and in industrial boilers. Beneficial use of CCPs refers to the use or substitution of CCPs for other products based on performance criteria. Under the RCC, EPA has established three goals for increased beneficial use of CCPs:

- Achieve a 50 percent beneficial use rate of CCPs by 2011;
- Increase the use of coal fly ash in concrete by 50 percent (from 12.4 million tons per year in 2001 to 18.6 million tons by 2011); and
- Reduce greenhouse gas emissions from concrete production by approximately 5 million metric tons CO_2 equivalent by 2010.[8]

Additionally, to support efforts to increase the beneficial use of CCPs, EPA has established partnerships with several industry groups and government agencies, including the American Coal Ash Association (ACAA), Utility Solid Waste Group (US WAG), the U.S. Department of Energy (DOE), and the Federal Highway Administration (FHWA). Efforts within these partnerships include: promoting the beneficial use of CCPs through the development of web resources; developing technical guidance on the best practices for the beneficial use of CCPs; holding educational workshops and outreach support for CCP users; and providing recognition for the innovative beneficial use of CCPs.

Overview of Report

This chapter serves two purposes: (1) To provide an initial assessment of the market dynamics that affect the generation, disposal, recovery, and beneficial use of CCPs; and (2) to provide a preliminary life cycle analysis of beneficial impacts of CCP use, including an initial estimate of the baseline beneficial use impacts with current (2005) CCP levels and, for some materials, the beneficial impacts associated with achieving the 2011 RCC goal. Ultimately, in combination with specific information about explicit RCC efforts, this chapter can be used to support the development and implementation of measures of program efficiency.

Organization of Report

The report proceeds in four chapters following this introduction. To provide market context, the second chapter characterizes the current generation and management of CCPs. The third chapter summarizes the current market structure for CCPs and outlines specific EPA efforts to increase their beneficial use. The fourth chapter uses baseline and Agency goal information, and available LCA tools to provide a preliminary life cycle analysis of the impacts of beneficial use of FGD gypsum and fly ash. The final chapter discusses the potential to extrapolate these beneficial use impacts and attribute them to EPA program efforts.

2. Baseline Characterization of CCP Generation and Beneficial Use

The coal-fired power industry is the largest generator of CCPs. Other industries, such as commercial boilers and mineral and grain processors that use coal as a fuel source also produce small quantities of CCPs. Because these other industries generate such small quantities of CCPs relative to the coal-fired electric power industry, this chapter focuses solely on the coal-fired electric power industry.[9]

CCPs are categorized by the process in which they are generated, which varies by plant. CCPs include the following materials:

- **Fly ash.** Exhaust gases leaving the combustion chamber of a power plant entrain fly ash particles during the coal combustion process. To prevent fly ash from entering the atmosphere, power plants use various collection devices to remove it from the gases that are leaving the stack. Fly ash is the finest of coal ash particles. The American Society for Testing and Material (ASTM) identifies two classes of fly ash suitable for beneficial use based on chemical composition. Class F fly ash results from the burning of anthracite or bituminous coal, while Class C fly ash results from the burning of lignite or subbituminous coal.
- **Bottom ash.** With grain sizes ranging from fine sand to fine gravel, bottom ash is coarser than fly ash. Utilities collect bottom ash from the floor of coal burning furnaces used in the generation of steam, the production of electric power, or both. The physical characteristics of the product generated depend on the characteristics of the furnace.
- **Flue Gas Desulphurization (FGD) material.** FGD material results from the flue gas desulphurization scrubbing process that transforms gaseous SO_2, released during coal combustion, to sulfur compounds. Coal-fired power plants employ either a wet or dry scrubbing method to remove SO_2 from their emissions. The final by-product of wet scrubbing is primarily FGD gypsum, although small amounts of other materials (e.g., ash, metals) are also produced.[10] In this chapter, we refer to these other materials as "other FGD wet material." The dry method produces by-products that consist of mainly calcium sulfite, fly ash, portlandite, and calcite. Collectively, we refer to these materials as "FGD dry material."[11] All three materials, FGD gypsum, other FGD wet scrubber material, and FGD dry scrubber material, can be used in a growing number of beneficial use applications.
- **Boiler Slag.** Boiler slag consists of molten ash collected at the base of cyclone boilers. Facilities cool boiler slag with water, which then shatters into black, angular pieces that have a smooth appearance.
- **Fluidized Bed Combustion (FBC) ash** (not pictured in Exhibit 2-1). A fluidized bed combustion boiler, a type of coal boiler that combines the coal combustion and flue gas desulphurization processes within a single furnace, generates FBC ash. FBC ash is rich in lime and sulfur.
- **Cenospheres.** Generated as a component of fly ash in high temperature coal combustion, cenospheres consist of extremely small, lightweight, inert, hollow spheres comprised largely of silica and alumina that are filled with low-pressure

gases.[12] When fly ash is disposed in settlement lagoons, cenospheres can be collected on the surface where they can be skimmed for use in manufacturing processes.

At a typical coal-fired power plant, coal combustion generates CCPs during several phases of the process. Exhibit 2-1 illustrates the collection of several types of CCPs. As depicted below, facilities remove bottom ash and boiler slag from the base of the furnace. Fly ash accumulates in the particulate collection device, while FGD material collects in the SO_2 control device.

Current Quantities of CCPs Generated and Managed

In 2005, the coal-fueled electric power industry generated approximately 123 million short tons of CCPs. Of these, the industry disposed of approximately 74 million short tons to landfills, while beneficially using approximately 50 million short tons in products. Exhibit 2-2, below, presents the current quantities of CCPs generated and managed, in the context of other materials targeted by the RCC. Except for construction and demolition material, the U.S. generates larger quantities of CCPs than other industrial and municipal solid waste (MSW).

The American Coal Ash Association (ACAA), a trade association whose purpose is to advance the beneficial use of CCPs, conducts an annual survey of coal-fired electric plants to collect data on the production, disposal, and use of CCPs in the U.S.[13] Exhibit 2-3 summarizes the 2005 survey on generation, disposal, and beneficial use of various CCP categories.

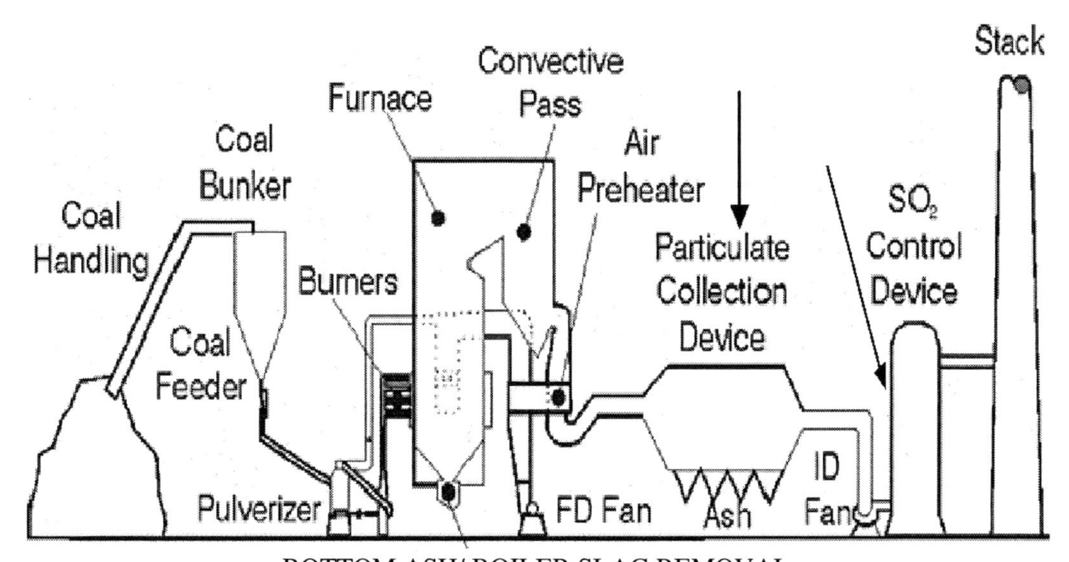

BOTTOM ASH/ BOILER SLAG REMOVAL

Source: Energy Information Administration, accessed at: www. eia. doe.gov.

Exhibit 2-1. Coal Combustion Process at a Coal-Fired Power Plant

Exhibit 2-2. RCC Materials by Quantitiy

MATERIAL[a]	QUANTITY GENERATED (MILLION SHORT TONS)	QUANTITY RECOVERED/ BENEFICIALLY USED[b] (MILLION SHORT TONS)	QUANTITY DISPOSED (MILLION SHORT TONS)	YEAR
C&D Material[1]	331	214[c]	118	2003
CCPs[2]	**123**	**50**	**74**	**2005**
Paper and Paperboard[3]	83	40	43	2003
Packaging[3]	74	29	45	2003
Organics[3]	56	17	39	2003
Foundry Sand[4, d]	9.2	2.6	6.6	2005
Chemicals[5]	0.04	NA	NA	2003

Notes:

a. Under the RCC 2005 Action Plan, increases in the rate of MSW recovery and reduction of priority and toxic chemicals are also targeted. We have included these material streams in this exhibit even though they are not targeted specifically for beneficial use.

b. The figures shown for paper and paperboard, packaging, and organics are the quantities recovered from the MSW stream. The figures shown for C&D debris, CCPs, and foundry sand are quantities that are beneficially used.

c. A Construction Materials Recycling Association member survey estimates that approximately 270 million tons of C&D material including asphalt and concrete from roads, bridge-related infrastructure, and land clearing debris was recovered in 2004.

d. The foundry sand quantity generated is uncertain, but estimates fall within the range of 6 to 10 million tons/year. Due to the lack of precise data on annual quantities generated and managed, the quantity disposed may include foundry sand that is being beneficially used as daily landfill cover.

Sources:

1. US EPA, "Characterization of Building-Related Construction and Demolition Debris in the United States" and "Characterization of Road-related Construction and Demolition Debris in the United States," 2005. (Note that these documents are preliminary and are currently undergoing peer-review).
2. American Coal Ash Association (ACAA), "2005 Coal Combustion Product (CCP) Production and Use Survey," accessed on October 29, 2006 at: http://www.acaa-usa.org/PDF/2005_CCP_Production_and_Use_Figures_Released_by_ACAA.pdf.
3. US EPA, "Municipal Solid Waste in the United States: 2003 Data Tables," Table 1, accessed on October 26, 2006 at: <http://www.epa.gov/epaoswer/non-hw/muncpl/pubs/03data.pdf>.
4. American Foundry Society (AFS). "Foundry Industry Benchmarking Survey," August 2007.
5. US EPA, "Draft National Priority Trends Report (1999-2003) Fall 2005," as reported in the NPEP GPRA 2008 database of TRI data from 1998-2003.

Exhibit 2-3 illustrates several important aspects of the generation, beneficial use, and disposal of CPPs:

- Of reported materials, fly ash constitutes the largest proportion (58 percent) of CCP materials generated in 2005. FGD material follows at 26 percent.[14] Bottom ash, boiler slag and FBC ash collectively comprise the remaining 17 percent of CCPs generated in 2005.

- Boiler slag and FGD gypsum have the highest percentage of beneficial use of the six coal combustion products.
- Fly ash, FGD material (other than FGD gypsum), and bottom ash have the highest disposal rates.

In addition to quantities of fly ash reported in the ACAA survey, stockpiles may provide another potential source of fly ash for certain beneficial uses. Industry sources estimate that between 100 million and 500 million tons of fly ash have accumulated in U.S. landfills since the 1920s, when disposal of fly ash in landfills began.[15, 16]

Exhibit 2-3. Summary of CCP Generation and Management in 2005

PRODUCT	CCPS GENERATED (MILLION SHORT TONS)	BENEFICIALLY USED (MILLION SHORT TONS)	PERCENT USED	QUANTITY DISPOSED (MILLION SHORT TONS)[a]	PERCENT DISPOSED
Fly Ash	71.10	29.12	41%	41.98	59%
Flue Gas Desulfurization (FGD) Material	31.10	10.12	33%	20.99	67%
Other FGD Wet Material	*17.70*	*0.69*	*4%*	*17.01*	*96%*
FGD Gypsum	*11.98*	*9.30*	*77%*	*2.71*	*23%*
FGD Dry Material	*1.43*	*0.16*	*11%*	*1.27*	*89%*
Bottom Ash	17.60	7.52	43%	10.06	57%
Boiler Slag	1.96	1.90	97%	0.07	3%
Fluidized Bed Combustion (FBC) Ash	1.37	0.94	69%	0.42	31%
Cenospheres[b]	Not available	0.08 [c]	Not available	Not available	Not available
Total CCPs	**123.13**	**49.61**	**40%** (see note 2)	**73.51**	**60%**

Notes:

a. Calculated by subtracting quantity beneficially used from quantity generated.

b. The ACAA's "CCP Production and Use Survey" does not report total generation or disposal quantities for cenospheres, only sales.

c. Follow-up communication with D. Goss on 11-10-07 indicated that this figure may be misreported in the 2005 CCP Survey. The actual figure is likely to be an order of magnitude less, or approximately 0.008 million short tons.

Note 2: Results from the 2006 CCP Production and Use Survey conducted by the ACAA indicate a total utilization rate of 43.43 percent, up from 40.29 percent reported for 2005. This reflects an ongoing upward trend in the CCP utilization rate over the past decade. The 2006 results were received too late for incorporation into the benefits analysis.

Source: American Coal Ash Association. "2005 Coal Combustion Product (CCP) Production and Use Survey," accessed at: http://www.acaausa. org/PDF/2005_CCP_Production_and_Use_ Figures _Released_by_ACAA.pdf.

Exhibit 2-4. Common Beneficial Uses for CCPs

CCP	BENEFICIAL USE
Fly Ash	**Concrete:** Concrete consists of a mixture of approximately 25% fine aggregate (sand), 45% gravel, 15% portland cement, and 15% water. Class C and class F fly ash can replace a percentage of the portland cement component of concrete. Fly ash contributes to enhanced concrete strength and durability, and is typically less expensive than portland cement.
	Cement clinker: Clinker is an intermediary product of the portland cement manufacturing process. Clinker is formed when a raw mix consisting of limestone, clay, bauxite, iron ore and quartz are heated in a kilm at higher temperatures. Fly ash can be blended with limestone or shale and fed into the cement kiln to make clinker, which is then ground into portland cement.
	Structural fill: Structural fill is an engineered material used to raise or change the surface contour of an area and to provide ground support beneath building foundations. It can also be used to form embankments. Depending on the soil type, fly ash can replace a percentage (generally 50 percent) of virgin rock, dirt, sand or gravel in structural fill.
	Waste stabilization: Fly ash can be used in place of portland cement, cement kiln dust, or lime to solidify and harden wet or liquid waste before it is landfilled. Class C fly ash hardens by itself in contact with moisture, but class F fly ash must be mixed with another hardening agent, such as portland cement, in order to be used in waste stabilization.
FGD Gypsum	**Wallboard:** Gypsum wallboard (or drywall) is used as an interior finish in the con-struction of homes and building. Wallboard is comprised of a layer of gypsum stu-cco sandwiched between two sheets of heavy paper. FGD gypsum can replace 100 percent of virgin gypsum in wallboard after the excess moisture has been removed.
	Agricultural soil amendment: FGD gypsum can be used to replace liming agents as an agricultural soil amendment for specific soil and crop types.
	Cement additive: In the production of portland cement, clinker is blended with a small amount of gypsum prior to grinding into finished portland cement. FGD gypsum can be used to offset virgin gypsum in cement manufacture.
Bottom Ash	**Structural fill:** Structural fill is an engineered material used to raise or change the surface contour of an area and to provide ground support beneath building foundations. It can also be used to form embankments. Bottom ash can be used to offset virgin sand and gravel in structural fill.
	Road base: A road base is a foundation layer underlying a pavement and overlaying a subgrade of natural soil or embankment fill material. It protects the underlying soil from the detrimental effects of weather conditions and from the stresses and strains induced by traffic loads. Bottom ash can be used to offset virgin sand or gravel in road base.
	Concrete: Bottom ash can be used as a coarse aggregate for concrete blocks, with its porous nature often qualifying the product for lightweight classification.
Boiler Slag	**Blasting Grit:** Blasting grit is an industrial abrasive used to shape, cut, sharpen, or finish a variety of other surfaces and materials. Boiler slag can be used as a replacement for other slags or virgin sand as blasting grit.
	Structural fill: Structural fill is an engineered material used to raise or change the surface contour of an area and to provide ground support beneath building foundations. It can also be used to form embankments. Boiler slag is occasionally used to offset virgin sand and gravel in structural fill.

Beneficial Use Options

The chemical and physical properties of CCPs allow for their use in a wide range of products. CCPs may be used as a component of various building materials (i.e., as a replacement for portland cement in concrete) or as a direct replacement for other virgin materials such as sand, gravel, or gypsum. The physical properties of CCPs make them especially useful for construction and industrial materials. Size, shape, and chemical composition determine the suitability of specific material flows for beneficial use. Higher value applications, such as use in cement or concrete products, require comparatively stringent specifications (in terms of size, shape and chemical composition), whereas lower value uses, such as structural or mining fills, can accept more variable materials. For this reason, EPA has found that lower technology applications that require large volumes of CCPs may present the greatest potential for expanded beneficial use.[17]

Exhibit 2-4 summarizes the most common beneficial uses for each CCP. As shown, this table excludes cenospheres and FBC ash as data on the primary beneficial uses of these materials are not available.

Exhibit 2-5, below, illustrates the quantities of CCPs being used in the most common beneficial use applications. The applications highlighted in the exhibit represent approximately 80% of the current use of CCPs.[18] We include an expanded version of this table, which details a more inclusive set of CCP beneficial use applications, in Appendix A.

Exhibit 2-5 illustrates several important aspects regarding the beneficial use options for CCPs:

- Concrete, wallboard, structural fill, cement, and waste stabilization comprise the highest volume beneficial uses of CCPs.
- The use of fly ash as a pozzolanic binder in concrete represents the largest single beneficial use application of a CCP material.[19] Fly ash can substitute for finished portland cement in concrete and can be a valuable additive to concrete mixtures that enhances the strength, durability, and workability of the concrete product.[20]
- FGD gypsum serves as a substitute for virgin gypsum in wallboard construction. This high value use represents the second largest use of CCPs, by volume, and second highest utilization rate at 77 percent.
- Although one of the smaller material streams, facilities beneficially use boiler slag in blasting grit, structural fill and waste stabilization, at the highest percentage of all CCPs. Boiler slag possesses two key properties that make it ideal for beneficial use: (1) the highly uniform quality of boiler slag increases its acceptance among potential end-users; and (2) boiler slag's unique abrasive properties make and excellent material for blasting grit and asphalt shingles.[21]

In comparison to the same ACAA survey conducted in 2004, total CCP utilization from 2004 to 2005 has increased slightly (0.21 percent). However, it is important to note that both the generation and beneficial use of CCPs increased during this time period. Both bottom ash and wet FGD material saw modest decreases in beneficial use rates (4% and 3%, respectively). The greatest increase in utilization rates over this time period was in boiler slag, with an increase of seven percent.[22, 23]

Exhibit 2-5. Key Beneficial Uses for CCPs in 2005 (Million Short Tons)

APPLICATION (INDUSTRY)	COAL FLY ASH	BOTTOM ASH	FGD GYPSUM	OTHER FGD WET MATERIAL	FGD DRY MATERIAL	BOILER SLAG	FBC ASH	TOTAL
Concrete[a] (Construction)	14.99	1.02	0.33	0	0.01	0	0	**16.35**
Structural fill[b] (Construction)	5.71	2.32	0	0	< 0.01	0.18	0.14	**8.35**
Wallboard[c] (Construction)	0	0	8.18	0	0	0	0	**8.18**
Raw feed for cement clinker[d] (Construction)	2.83	0.94	0.40	< 0.01	0	0.04	0	**4.22**
Waste stabilization[e] (Waste Mgmt)	2.66	0.04	0	0	0	0	0.14	**2.84**
Blasting Grit/Roofing Granules	0	0.89	0	0	0	1.54	0	**1.63**
Total - Key Uses	***26.19***	***4.41***	***8.90***	***< 0.01***	***0.02***	***1.76***	***0.28***	***41.57***
Total – Other Uses[f]	***2.93***	***3.13***	***0.36***	***0.69***	***0.014***	***0.13***	***0.66***	***8.04***
TOTAL - ALL USES	**29.12**	**7.54**	**9.27**	**0.69**	**0.16**	**1.89**	**0.94**	**49.61**
2005 QUANTITY GENERATED	**71.10**	**17.60**	**12.00**	**17.70**	**1.43**	**1.96**	**1.37**	**123.13[g]**
CCP UTILIZATION RATE[h]	**41%**	**43%**	**77%**	**4%**	**11%**	**97%**	**69%**	**40%** (see note 2)

Notes:

a. CCPs are frequently used as a replacement for a portion of portland cement in the manufacture of concrete.

b. Structural fill is an engineered material that is used to raise or change the surface contour of an area and to provide ground support beneath highway roadbeds, pavements and building foundations. It can also be used to form embankments.

c. FGD gypsum is used as a substitute for virgin gypsum in wallboard manufacturing.

d. CCPs can be blended with limestone or shale and fed into the cement kiln to make clinker, which is then ground into portland cement.

e. The chemical properties of CCPs make them effective stabilizers of biosolids (i.e., sludge from municipal waste water treatment).

f. Includes quantities beneficially used in minor applications not included in this exhibit, but listed in Appendix A.

g. Includes 115,596 tons of "Other FGD Material" not listed in this table because of the small quantities generated.

h. CCP utilization rates reflect all use applications, some of which are omitted from this table but are included in Appendix A. Utilization rates are calculated by dividing the total quantity used by the total quantity generated.

Note 2: Results from the 2006 CCP Production and Use Survey conducted by the ACAA indicate a total utilization rate of 43.43 percent, up from 40.29 percent reported for 2005. This reflects an ongoing upward trend in the CCP utilization rate over the past decade. The 2006 results were received too late for incorporation into the benefits analysis.

Sources:

1. American Coal Ash Association. "2005 Coal Combustion Product (CCP) Production and Use Survey," accessed at: http://www.acaa-usa.org/PDF/20045_CCP_Production_and_ Use_Figures_ Released_by_ACAA.pdf.
2. Western Region Ash Group, "Applications and Competing Materials, Coal Combustion Byproducts," accessed at: http://www.wrashg.org/compmat.htm.

3. MARKET STRUCTURE OF BENEFICIAL USE FOR CCPS

Understanding the factors that affect beneficial use of CCPs requires consideration of the underlying markets affecting its generation and management. The CCP market includes three market segments: (1) coal-fired utilities, (2) intermediaries: CCP marketers and consultants, and (3) end-users. In addition, state regulators determine the extent to which CCPs can be beneficially used by defining the regulatory context in which these actors operate. This chapter considers the factors affecting beneficial use decisions among these groups participating in the marketplace. We then present a discussion of opportunities for growth in the general CCP markets, along with a more specific illustration using three common beneficial applications. Finally, we discuss our efforts to improve market conditions for CCPs.

Three main challenges exist in developing the beneficial use market for CCPs. First, because CCPs are a heavy material to transport, the distance between the location of the coal-fired utility generating the CCPs and the potential end-user is a driving factor in determining whether the CCPs will be beneficially used in a project. Another difficulty in developing the beneficial use market is the capacity for individual coal-fired utilities to provide a quantity of high quality CCPs sufficient to meet the end-users' demands. The ability of end-users to obtain enough CCPs for their purposes is an important consideration in driving demand for CCPs. Finally, as noted above, the variability in use options across states poses a challenge to both coal-burning plants and end-users in trying to determine applicable beneficial use options for CCPs.

Coal-Fired Utility Practices: CCP Supply

The coal-fired power industry is the largest generator of CCPs in the United States. As noted previously, other industries that use coal as a fuel source in commercial or industrial boilers (e.g., mineral and grain processors) also produce small quantities of CCPs. Coal-generated electricity supplies approximately 50% of the electricity consumed in the United States.[24] Since electricity demand is projected to increase by 40% by 2020 and coal will continue to be an important fuel source, it is likely that the quantity of CCPs produced and available for beneficial use will also increase.[25, 26]

Approximately 400 to 500 coal-fired electric utilities currently operate in the U.S.[27] Exhibit 3-1 shows the geographic distribution of coal consumption by electric power plants across the U.S. Coal consumption by power plants is greatest in the East North Central region of the U.S., but consumption remains relatively high throughout the entire Central and Southern United States. Coal consumption is low in the contiguous and noncontiguous Pacific regions of the U.S. and in New England. CCP generation closely approximates the geographic distribution of coal consumption across the U.S., but CCP generation is not directly proportional to coal consumption. The composition of coal varies regionally in the U.S. For example, the non-combustible portion (commonly referred to as "ash") of Western bituminous coal is higher than that of Western sub-bituminous coal (approximately 10% to 15% and 4% to 6% ash, respectively). Coal with a higher non-combustible ash content will yield greater quantities of CCPs when combusted.

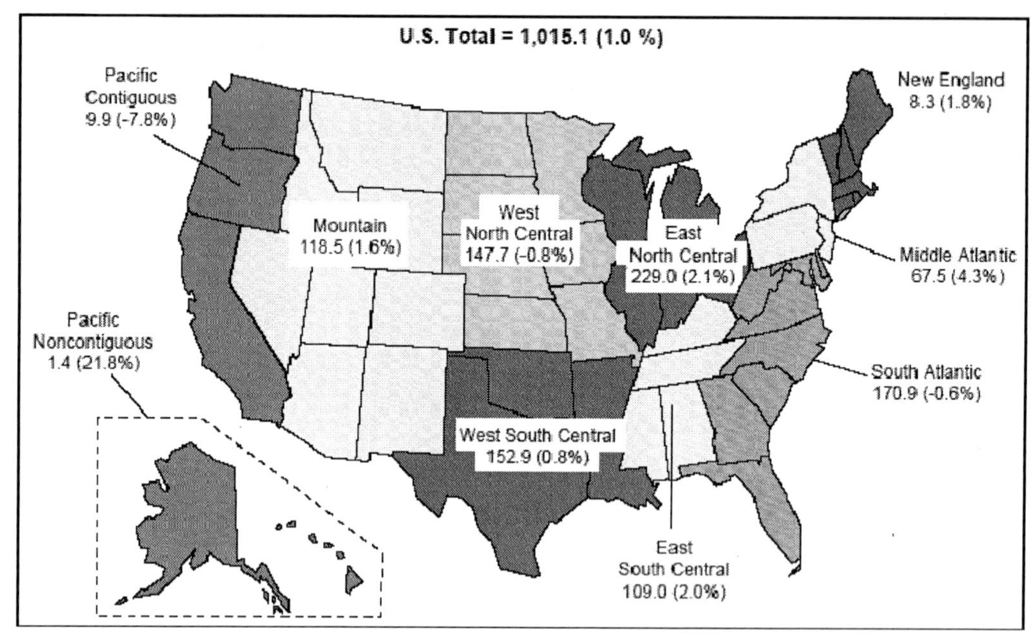

Source: Energy Information Administration, accessed at: www.eia.doe.gov.

Exhibit 3-1. Electric Power Sector Consumption of Coal in 2004, by Census Region (Million Short Tons and Percent Change from 2003)

Several factors influence a utility's decision to supply CCPs for beneficial use. Economic factors are the primary consideration and include:

- **Landfill disposal costs.** For many utilities, the sale of CCPs for beneficial use is a means of reducing operating costs through avoidance of landfill tipping fees. In order for beneficial use of CCPs to be competitive, the cost of reselling CCPs, minus revenue from the sale, must be less than the cost of landfill disposal. Landfill tipping fees vary regionally but range from $5 per ton to $45 per ton.[28] Avoiding landfill disposal costs may be a significant incentive for a utility to engage in beneficial use.
- **Revenue from sale.** Depending on the type of CCP, an electric utility may or may not receive revenue for its ash. For some CCP types, marketers will accept ash as a service to the plant (allowing the plant to avoid disposal costs) but do not pay for the ash. For other CCP types, especially fly ash, boiler slag and cenospheres, the revenue received can be a significant incentive for a utility to market its ash.[29]
- **Transport costs.** CCPs are heavy materials, which makes transport over long distances expensive. Transport distance between the utility and the nearest landfill or end-user is a significant determinant in the management of CCPs.
- **Processing costs.** Approximately 90% to 95% of CCPs do not require processing prior to beneficial use. However, higher value applications that require specialized CCP products may require processing to meet material specifications.
- **Storage costs.** In many parts of the country, the production of coal ash is high during both the coldest and hottest months of the year when people are heating and cooling their homes, offices and schools. However, the winter season is often the slowest

period for construction and other applications that beneficially use the fly ash. As a result, it is necessary to store CCPs until they can be utilized. Typically, domes are inflated adjacent to boilers for the CCP collection. The cost of storing fly ash or other CCPs during the winter months may be a deterrent to beneficial use by a utility.

- **Marketing costs.** In order to attract buyers of CCPs, a marketer must devote financial resources to marketing their CCP product(s) for beneficial use. Some utilities market their CCPs directly to end-users, but others pay a third party marketer or broker to negotiate CCP sales.

In the end, a generator's decision to make CCPs available for beneficial use rather than disposal is a result of a confluence of all the above factors, as well as non-economic factors such as access to information about permissible applications and availability of technical assistance. Depending on the circumstances, a coal-fired utility may weigh certain factors more heavily than others. For example, since off-site disposal and transport to a marketer or end-user both require hauling, total transport distance to the off-site facility is an important consideration. If the off-site disposal facility is closer, then a generator may opt to send its CCPs to the landfill instead of to the marketer or end-user. However, if the avoided disposal costs from the marketing arrangement are higher than the cost of offsite disposal, then avoided costs may be enough to offset the additional cost to transport the materials to the marketer or end-user. In addition, a coal-fired utility may be more willing to absorb higher costs (of transport, marketing, etc.) for higher- value materials and uses for which it can charge a higher premium (e.g. fly ash as a portland cement substitute in concrete).[30]

Intermediaries

Many coal-fired electric generators market their CCPs through a third-party marketer instead of selling directly to the end-user. In these cases, a utility perceives an efficiency in outsourcing the marketing of its CCPs. Marketers typically accept all of a generator's CCPs as a service to the company, sell the marketable portion, and dispose of the portion that is not salable. The marketer typically bears the cost of hauling CCPs from the utility and incorporates this cost into the sale price.

End-Users and Purchasers: CCP Demand

Several factors influence an end-user's decision to use CCPs in their product. Such considerations include:

- **Price of CCPs relative to the price of virgin materials.**[31] If the price of a virgin material is less than the price of CCPs (which will reflect cost factors such as transport distance, processing and storage costs), end-users will generally purchase virgin materials. In areas where virgin materials are abundant and inexpensive, CCPs may not be economically viable. Exhibit 3-2 shows the typical price ranges for CCPs used in various applications relative to the virgin materials they replaces.

Exhibit 3-2. Sample CCP and Virgin Material Prices for CCP Applications

VIRGIN MATERIAL	2005 AVG PRICE, (PER TON, FREE ON BOARD)[a,b]	CCP SUBSTITUTE	2005 AVG PRICE, (PER TON, FREE ON BOARD)[a]
Portland cement	$80	Concrete quality fly ash	$0 to $45
		Boiler slag	Not available
Virgin aggregate for fill	$3	Fly ash for flowable fill	$1
Virgin aggregate for road base	$5	Bottom ash or fly ash for road base	$4 to $8
Lime for soil stabilization (Hydrated lime)	$83	Fly ash for soil stabilization	$10 to $20
Lime for waste stabilization (Quicklime)	$66	Fly ash for waste stabilization	$15 to $25
Virgin aggregate for snow and ice control	$5	Bottom ash for snow and ice control	$3 to $6
Gypsum for wallboard interior	$4.50 - $12.0	FGD Gypsum	$0 - $8.00

Notes:

a. Virgin material prices are reported by USGS while CCP prices are provided by ACAA. This price data represents the best available information, and should be cross-compared with caution, as the data may not capture all factors driving price variability.

b. "Free on Board" is a shipping term, which indicates that the supplier pays the shipping costs (and usually the insurance costs) from the point of manufacture to a specified destination, at which point the buyer takes responsibility.

Sources:

1. USGS, "Mineral Commodities Summary 2006: Cement," accessed at: http://minerals.usgs.gov /minerals/pubs/commodity/cement/cemenmcs06.pdf
2. USGS, "Mineral Commodities Summary 2004: Construction Sand and Gravel," accessed at: http://minerals.usgs.gov/minerals/pubs/commodity/sand_&_gravel_construction/sandgmyb04.pdf
3. USGS, "Mineral Commodities Summary 2005: Lime," accessed at: http://minerals.usgs.gov/minerals /pubs/commodity/lime/lime_myb05.pdf
4. USGS, "Mineral Commodities Summary 2006: Gypsum," accessed at: http://minerals.usgs.gov /minerals/pubs/commodity/gypsum/gypsumcs06.pdf
5. American Coal Ash Association, "Frequently Asked Questions," accessed at: www.acaa-usa.org
6. Miller, Cheri . Gypsum Parameters. Presentation at WOCA Short Course: "Strategies for Development of FGD Gypsum Resources."

- **Technical fit between CCPs and use application.** CCPs have varying physical and chemical characteristics due to differences in coal types, combustion processes, air pollution control technologies, and CCP management practices at individual power plants. To be beneficially used in a particular application, the chemical and physical properties of the CCP must align with the engineering requirements of that application. For example, high carbon content or the presence of air emission additives may render some CCPs unsuitable for some use applications.

- **Sufficient quantities of CCPs.** Some beneficial use applications require larger volumes of CCPs than are typically produced at a single power plant. Where demand for CCPs is greater than the supply generated by a single plant, the end-user may need to purchase CCPs from multiple suppliers; this can increase transaction costs.
- **State Regulations.** Regulations governing beneficial use of CCPs vary by state. In many states, beneficial use of CCPs must be approved on a project-by-project basis. Currently, public and environmental health considerations drive state regulatory decisions concerning beneficial use of CCPs in end-use applications.[32,33]
- **Incomplete science.** In absence of definitive data on health risks associated with the beneficial use of CCPs, some states have chosen to limit the use of CCPs in building materials. For example, EPA research has found that CCPs may release small quantities of mercury to the ambient air during use in certain industrial processes.[34] Noting this research, States have questioned the safety of using fly ash in cement to be used in schools.[35]

Market Dynamics of Specific Use Applications

The beneficial use markets for CCPs depend on the physical properties of the materials, the demand for their particular uses, and the supply of materials available for use. Exhibit 3-3, below, summarizes the current state of the beneficial use markets for the suite of CCPs, along with an analysis of the potential for growth in the beneficial use market for each material. An established market contains four key elements, all of which are interconnected:

- Generators producing a consistent supply of materials;
- End-users with the demand to use the product;
- Well-known, accepted beneficial use applications; and
- A distribution system to transport materials from generators to users.

Limited markets may have a subset of these key elements, but likely need a shift in technology, demand, or price to increase beneficial use of the product. Emerging markets are typically unorganized and the connections between the elements are not fully formed.

Aside from the boiler slag market, the markets for the various CCPs generally have room for growth. EPA programs that aim to increase growth in the markets for CCPs by targeting beneficial use applications that are well-accepted practices in the industry may have significant success in helping expand these markets by overcoming targeted technical, administrative and technical and informational hurdles. The economic viability of the top three beneficial uses (by volume) is considered individually below. Concrete, gypsum wallboard, and structural fill are all long-standing, widely accepted uses for CCPs.

Concrete

Certain performance benefits can be attained through the use of coal fly ash in concrete, including greater workability, higher strength, and increased longevity in the finished concrete product.[36] Fly ash substitutes directly for portland cement in the concrete mixing process. This beneficial use represents one of the highest value applications for CCPs, and has the potential to increase as a result of the current high demand for portland cement in the U.S.[37,38]

Exhibit 3-3. CCP Beneficial Use Markets

CCP	STATE OF BENEFICIAL USE MARKET	POTENTIAL FOR MARKET GROWTH
Fly Ash	Established markets	The markets for fly ash have the potential to continue to grow. Generators produce fly ash in large quantities and there are several well-known, high-value uses. Currently, only 41% of fly ash is beneficially used.
FGD Gypsum	Established markets	The FGD gypsum market has room for moderate growth. FDG gypsum wallboard is an accepted alternative to virgin gypsum wallbaord with users often directly connected to generators. Currently, 77% of FDG gypsum is beneficially used. Other uses are emerging but are currently limited, and data on market opportunities are limited.
Dry FGD Material	Limited market	The current market for dry FDG material appears to have only limited potential for growth. Uses for the material appear to be limited to relatively low-value uses, such mine reclamation. Currently, only four percent of this material is beneficially used. Data on the market opportunities for this material are limited.
Other Wet FGD Material	Limited market	The current market for other FDG wet material appears to have only limited potential for growth. Uses for the material appear to be limited to relatively low-value uses, such mine reclamation. Currently, only 11% of this material is beneficially used. Data on the market opportunities for this material are limited.
Bottom Ash	Established markets	The markets for bottom ash have the potential to continue to grow. Generators produce bottom ash in large quantities and there are several well-known, high- value uses. Currently, only 43% of bottom ash is beneficially used.
Boiler Slag	Established markets	The markets for boiler slag are mature and have limited opportunity for growth. Since approximately 97% of boiler slag is beneficially used (primarily as blasting grit), supply is currently roughly equivalent to demand.
FBC Ash	Established markets	The markets for FBC ash have potential for moderate growth, although mainly for low-value uses. Currently nearly 70% of FBC ash is beneficially used.
Cenospheres	Emerging market	The markets for cenospheres have the potential for growth as information on their uses becomes more widely available. However, information regarding potential uses is limited at this time. Currently the beneficial use rate for cenospheres is unknown.

Industry representatives and some state agencies have stated that there are possible emerging issues related to regulatory programs for the control of nitrogen oxides (NOx) and mercury in power plant flue gases. These issues relate to the potential for negative impacts on coal fly ash quality and available quantities due to the potential for increased mercury contamination in the ash, or unacceptably high levels of carbon content. There are technology choices that would minimize these impacts on the beneficial use of coal fly ash. However, the selection of equipment for control of nitrogen oxides and mercury, and corresponding technologies potentially necessary to minimize quality impacts on coal fly ash is very complex, resulting in industry solutions that would be unit-specific. Losses of anywhere between $40/ton and $80/ton of coal fly ash[39] are possible if industry is unable to sell high carbon fly ash as a supplementary cementitious material in the manufacture of concrete. This estimate also includes the additional costs associated with the need to dispose of a formerly marketable material.

State of Florida officials noted that installation of air emission controls at coal-fired power plants might result in increased mercury associated with coal fly ash. Since coal fly ash is used in portland cement manufacturing, electric utilities and portland cement manufacturers have expressed concern that the Florida Department of Environmental Protection's (FDEP) New Source Review (NSR) might limit or even eliminate the use of CCPs for this purpose. Officials also noted that similar impacts might occur for coal fly ash containing higher levels of unburned carbon or other components resulting from changes in operations, fuel, or emission controls.[40] This may have the potential of jeopardizing the recycling of many tons of CCPs that are currently reused.

The Texas CCP review also notes that emissions control in the electric utility industry has had a subsequent impact on the type, quantity, and quality of the solid materials produced at a specific power plant[41]. Officials indicate that the reduced supply of high quality coal fly ash already poses a threat to coal fly ash use in TX DOT projects, where high volumes of consistent quality coal fly ash are needed over the duration of large, long-term projects.

Overall, technology options are available to the industry, specifically for the application of NOx controls, which would minimize any impacts on the quality of fly ash. Furthermore, technology solutions are currently being developed and deployed in the industry to minimize or avoid any such impacts from the use of mercury controls as well.

Gypsum Wallboard

FGD gypsum is a product derived from the wet FGD process. Utilization of FGD gypsum in wallboard manufacture is a well-established market. Because the quality of FGD gypsum produced by power plants is generally consistent, new wallboard facilities often locate adjacent to power plants to allow FGD gypsum to be delivered directly to the wallboard plants. In some cases, wet FGD gypsum is piped directly to the adjacent wallboard facility, a step that significantly reduces transport and handling costs. Given these developments, the demand for FGD gypsum will likely remain high, and may increase as new wallboard manufacturing facilities are being constructed to accommodate FGD gypsum in wallboard production.[42]

Structural Fill

Structural fill is an engineered material used to raise or change the surface contour of an area and to provide ground support beneath highway roadbeds, pavements, embankments and building foundations. The quality and engineering standards for use of CCPs in structural fill are less stringent than the standards for structural applications such as concrete. Consequently, CCPs destined for use in structural fill generally do not require processing, which keeps costs low.

Demand for CCPs in structural fill applications is variable and generally occurs on a project-by-project basis. One large construction project using CCPs in fill can create a spike in demand for CCPs, but this may be followed by a lull in demand until another sizeable project can be identified.[43] Because CCPs are generated continuously, the generator's or marketer's capacity to store and accumulate the material between projects is a significant determinant in the use of CCPs in structural fill.

Impacts of Current Policy Setting on Market Dynamics

While states play a primary role in establishing industrial waste regulations and guidance, EPA has an opportunity to provide coordination and assistance at the national and regional levels to help achieve a shift in waste management policy. EPA is currently engaged in several long-term efforts to increase beneficial use of CCPs.

Under the RCC, EPA established goals for beneficial use of CCPs (as enumerated in the introduction) and established the Coal Combustion Products Partnership (C^2P^2) to help reach these goals. C^2P^2 is a cooperative effort among EPA, ACAA, the Utility Solid Waste Activities Group (US WAG), the U.S. Department of Energy (DOE), the U.S. Federal Highway Administration (FHWA), the Agricultural Research Service of USDA, and the Electric Power Research Institute (EPRI). Through C^2P^2, EPA and its co-sponsors work with all levels of government, as well as industry organizations, to identify and address regulatory, institutional, economic, and other limiting factors to the beneficial use of CCPs. One important overarching barrier addressed by C^2P^2 is the lack of information about beneficial use opportunities. Specifically, the program includes the following initiatives and activities:

- **The C^2P^2 Challenge:** Under the C^2P^2 challenge, partners are eligible for awards recognizing activities such as documented increases in CCP use and successes in CCP promotion and utilization.
- **Barrier Breaking Activities:** C^2P^2 addresses limiting factors to increased CCP utilization through activities such as developing booklets and web resources on the benefits and impacts of using CCPs in highway and building construction applications; publishing case studies on successful beneficial use of CCPs; supporting Green Highways; and updating a manual for highway engineers on the use of fly ash in highway applications.
- **Technical Assistance:** C^2P^2 has conducted a series of workshops with FHWA, EPA, DOE, ACAA and other partners to provide technical assistance and outreach to support the use of CCPs in concrete highway construction. These workshops present

the technical feasibility of using CCPs and the economic and environmental benefits that result from their use.

4. IMPACTS ASSOCIATED WITH BENEFICIAL USE OF CCPS

To evaluate EPA's efforts to improve the beneficial use of CCPs, it is essential to quantify the important environmental and human health impacts of increased use of these materials in various beneficial use applications. An initial step in this process is describing the incremental environmental impacts associated with using a specific quantity (e.g., one ton) of CCPs in different applications. These impacts can then be extrapolated in specific scenarios designed to address program-level outcomes. Life cycle analysis (LCA) represents a proven methodology for describing the impacts of beneficial use of specific quantities of material, and can also inform a broader evaluation of program achievements.

This chapter first provides a brief overview of the use of LCA in the assessment of environmental impacts of beneficial use, and also discusses the relationship between LCA and the economic analysis of net social benefits. Next, the chapter identifies several available LCA tools that can be used to provide insights into the impacts of beneficial use of CCPs in different applications, and presents an initial life cycle analysis of the potential impacts associated with the use of one ton of fly ash and FGD gypsum in concrete and wallboard manufacture, respectively. Finally, we note key limitations of this initial assessment and identify areas for additional research.

Life Cycle Analysis and RCC Program Outcomes

Life cycle analysis depicts the production of materials as a system of complex physical outcomes, and can predict the incremental physical consequences of a change in material inputs, technology, waste management practices, or price incentives. In LCA, as in reality, one change in the physical system, such as the substitution of fly ash for virgin portland cement, leads to a corresponding cascade of economy- wide impacts and shifts. As inputs are substituted, technologies, physical outputs, and exposure pathways change. Using a range of modeling platforms and life cycle inventories to calculate the outputs associated with each intermediate change, LCA calculates the net result of all of these interactions, capturing the total incremental effect of a change in operations on physical environmental impacts such as air emissions, and energy and water use. Life cycle analysis can be an effective performance assessment tool, and because it is a systems approach to assessment, it represents an improvement over less comprehensive techniques.

The RCC is designed to help facilitate changes in the economics and practice of waste generation, handling, and disposal (e.g., by promoting market opportunities for beneficial use). The outcomes of the program, therefore, can be described as the changes in total environmental impacts that result from changes in beneficial use. Many of these impacts likely come from avoiding the production of virgin materials that would be used in the absence of industrial materials. In some cases, changes in materials use may also lead to (positive and negative) changes in processing, product performance, and disposal approaches.

LCA can, given appropriate data and modeling scenarios, describe the net impacts of all of these changes, and can, therefore, provide an assessment of program results.

Life Cycle Analysis and Economic Benefit Assessment

As a tool for measuring physical impacts of system changes, LCA is a natural starting point in the assessment of the economic benefits of a program, but it is important to distinguish between LCA and economic benefits analysis. LCA is useful in the context of benefits analysis because it reflects a systems approach, allows measurement of changes to baseline conditions, identifies tradeoffs, and yields concrete, measurable metrics that can be evaluated both in isolation and comparatively, across programs and activities.

However, while it can provide a clear assessment of beneficial (and other) program *impacts*, LCA does not itself measure the social *benefits* and costs of changes in practice, for two reasons. First, LCA provides a *static* examination of impacts based on a one-time change to a system, and does not attempt to measure net impacts over time, adjusting for long-term market responses (e.g., changes in price and behavior) that can, in turn, affect the long-term system operation.[44] Beneficial use of large-volume CCP materials, such as fly ash and FGD gypsum, is already well-established. Therefore, gradual increases in use of these materials may have some impact on the large cement and gypsum markets in the U.S., but this analysis assumes that dramatic near-term changes in the market are unlikely. Therefore, the impacts estimated by LCA are reasonable representations of total market impacts. However, if sudden, large- scale changes in production of raw materials occur as a result of RCC efforts, then a net economic impact analysis using methods like partial equilibrium analysis might be necessary.

Second, a complete assessment of the net economic benefits of a program requires the application of economic valuation techniques to the physical outputs of LCA analysis, in order to describe, in economic terms, what the physical outcomes imply for human well-being. For example, an LCA can describe changes in the quantity of water used or waste generated in a process, but is not designed to identify the effect of these impacts on well-being.[45] Economic valuation of these changes depends upon the specific location, timing, and quality of the water that is consumed or waste disposal that occurs. The value of that water depends on how it would otherwise be used (e.g., for human consumption, industrial uses, habitat support, irrigation) and the "value" of the waste depends on the health risks it poses, release scenarios, and the people exposed.[46]

Ideally, LCA would be incorporated into a full-scale analysis of net program benefits that would account for market responses and would value specific environmental impacts such as decreased releases of pollutants. Unfortunately, the translation of physical changes into economic outcomes is typically costly, difficult, and often controversial when applied to human health or environmental outcomes. It frequently requires location-specific data on releases and exposures, as well as well-documented links between these exposures and health or environmental impacts. Assigning an economic value to even a small set of physical impacts can be a significant and expensive undertaking.[47]

Accordingly, LCA can represent not only a necessary ingredient, but also a practical initial alternative to a complete economic benefit assessment. While net economic benefits are often the target performance measure, it is necessary in many cases to rely on simpler proxies to facilitate management and performance assessment. As proxies, LCA outputs can represent a legitimate and defensible measure of program impacts. Therefore, we use LCA to investigate the measurable *beneficial impacts* associated with RCC program achievements, using available LCA tools. A more detailed discussion of the role of LCA in economic benefits assessment is provided in Appendix B.

Assessment of Beneficial Impacts of CCP Use

To fully capture the beneficial impacts of C^2P^2 program achievements, it would be necessary to model each beneficial use application of all CCPs targeted by the RCC. However, the time, data, and resources required to perform this task are beyond the scope of this chapter. For this preliminary analysis, we have selected two common CCPs, fly ash and FGD gypsum, which have well-understood beneficial use applications and processes, and for which life cycle models and existing data are available.

We conducted a comprehensive review of available data sources and tools for assessing life cycle benefits of beneficial use of these materials across all possible use applications. We identified four models that support specific evaluation of the environmental impacts of avoided virgin materials extraction in processes where CCPs are beneficially used in place of virgin materials:

- **BUILDING for Environmental and Economic Sustainability (BEES)** was developed by the National Institute of Standards and Technology (NIST) with support from the U.S. EPA to allow designers, builders, and product manufacturers to compare the life cycle environmental, and economic performance of alternative building products.[48] The BEES methodology measures environmental performance using an LCA approach, following guidance in the International Standards Organization 14040 series of standards for LCA. Thus, all stages in the life of the product are analyzed: raw material acquisition, manufacture, transportation, installation, use, and recycling and waste management. The BEES model is implemented in publicly available decision-support software, complete with actual environmental and economic performance data for a number of building products.
- **SimaPro** was developed by the Dutch company Pré Consultants and can be used to perform detailed lifecycle analyses of complex products and processes. SimaPro provides a high degree of modeling flexibility in that it provides data profiles representing thousands of materials production, transport, energy production, product use and waste management processes that can be combined to model very specific systems. Thus, SimaPro relies on the user's understanding of the various lifecycle stages and processes in the system being modeled. Results can be displayed as lifecycle inventory flows (e.g. energy use, water use and pollutant emissions (for a variety of pollutants including the criteria pollutants). In addition, one of several impact assessment methods can be applied to characterize the environmental damages (e.g., global warming, eutrophication, etc.) associated with these flows.
- **Pavement Life Cycle Assessment Tool for Environmental and Economic Effects (PaLATE)** is an Excel-based tool developed by the Consortium for Green Design and Manufacturing at U.C. Berkeley for life cycle analysis of environmental and economic performance of pavements and roads. The model was developed for

pavement designers and engineers, transportation agency decision-makers, civil engineers, and researchers. PaLATE can evaluate the relative impacts of using different virgin and secondary materials in the construction and maintenance of roads. Based on user-specified data on the type and quantity of initial construction materials, road construction equipment (e.g., asphalt paver), material transportation distances and modes, maintenance materials and processes, and off-site processing equipment (e.g., rock crusher), PaLATE calculates twelve life cycle inventory flows including water and energy use, conventional air emissions (NOx, SO_2, CO_2, PM_{10}, and CO), toxic air emissions (Pb and Hg), RCRA hazardous waste generation; and cancer and non-cancer Human Toxicity Potentials.[49]

- The **WAste Reduction Model (WARM)** was created by EPA to help solid waste planners and organizations estimate greenhouse gas (GHG) emission reductions from several different waste management practices. WARM calculates GHG emissions for baseline and alternative waste management practices, including source reduction, recycling, combustion, composting, and landfilling. The user can construct various scenarios by entering data on the amount of waste handled by material type and by management practice. WARM then automatically applies material-specific emission factors for each management practice to calculate the GHG emissions and energy savings of each scenario. In addition, the model will convert these outputs to equivalent metrics including the equivalent number of cars removed from the road in one year, the equivalent number of avoided barrels of oil burned, and the equivalent number of avoided gallons of gasoline consumed.

All four models support life cycle analysis of various CCPs. PaLATE, BEES and WARM all include life cycle data to evaluate use of fly ash as a substitute for finished portland cement in concrete, and SimaPro supports evaluation of use of FGD gypsum in wallboard. We select BEES and SimaPro to evaluate fly ash and FGD gypsum beneficial use because these models have been peer-reviewed and evaluate a large suite of environmental metrics.[50] In contrast, PaLATE has not undergone a formal peer review process, and WARM evaluates only greenhouse gas metrics. For comparative purposes, however, we present the results of a WARM analysis of the use of fly ash in concrete in Appendix C.[51]

In addition to BEES and SimaPro, the U.S. Economic Input-Output Life Cycle Assessment (EIO-LCA) model provides an alternative approach for measuring the avoided upstream impacts of recycling. EIOLCA was developed at Carnegie Mellon University and provides the capacity to evaluate economic and environmental effects across the supply chain for any of 491 industry sectors in the U.S. economy. EIOLCA also can represent the supply chain use of inputs and resulting environmental outputs across the supply chain by using publicly available data sources from the U.S. government. By integrating economic data on the existing flow of commerce between commodity sectors with environmental data on releases and material flows generated by each sector, it is possible to estimate the additional environmental emissions caused by an increase in production within a particular sector, accounting for changes in the supply chain. This approach can be used to provide insight into the sectors of the economy that drive the environmental impacts of a given process, and shed light on the specific impacts of particular policy efforts. While it is very helpful in examining the distribution of impacts across economic sectors, the EIO-LCA is not optimal for a specific life cycle analysis of beneficial use of FGD gypsum in wallboard because the life cycle impacts are modeled at the sector level and do not provide the same process-level resolution that can be estimated for various use applications using SimaPro. EIO-LCA is more useful for modeling use of fly ash in concrete, as it includes data for a fairly homogenous cement sector. It is also important to note that EIO-LCA is a dollar-based model and thus, is not directly comparable to the BEES/SimaPro data that is presented in tons. Appendix C describes supply chain manufacturing impacts for cement and gypsum production modeled using EIO-LCA.

Methodology for Evaluating Unit Impacts of Beneficial Use

We conduct separate analyses to evaluate the incremental environmental impacts associated with beneficially using a specific quantity (i.e., one ton) of fly ash and FGD gypsum. We employ:

- BEES to investigate using one ton of fly ash as a substitute for finished portland cement in concrete; and
- SimaPro to investigate using one ton of FGD gypsum as substitute for virgin gypsum in wallboard.

The first step in evaluating the life cycle impacts of beneficial use of fly ash and FGD gypsum is development of environmental impact profiles for use of one ton of each material as a substitute for portland cement in concrete and for virgin gypsum in wallboard, respectively. One ton was selected as the unit-basis for these analyses because the impacts can then easily be extrapolated to current use quantities, which are reported by the American Coal Ash Association (ACAA) in tons. In addition, by developing life cycle benefits profiles for use of a consistent quantity of each material, the impact profiles of specific materials can be compared with each other.

The calculation of unit impact values for fly ash and FGD gypsum are described in greater detail below. To the extent possible, we attempt to use comparable assumptions and life cycle system boundaries in both analyses.

BEES Analysis of Use of Fly Ash in Concrete

BEES includes environmental performance data for a number of concrete products (e.g., concrete columns, beams, walls, and slab on grade). The user can compare the environmental performance data of each of these products using different pre-determined concrete mix-designs, some of which include fly ash. The BEES environmental performance data serve as quantified estimates of the energy and resource flows going into the product and the releases to the environment coming from the product, summed across all stages of the product life cycle for one cubic yard of concrete. BEES quantifies these flows for hundreds of environmental metrics, but to capture the general spectrum of impacts, we focus on the following:

- Total primary energy use (MJ);
- Renewable energy use (MJ);
- Nonrenewable energy use (MJ);

- Water use (liters);
- Atmospheric emissions (CO_2, methane, CO, NOx, SOx, particulates, Hg, Pb) (grams);
- Waterborne waste (suspended matter, biological oxygen demand, chemical oxygen demand, Hg, Pb, selenium; and
- Nonhazardous waste (kg).

As an example of the LCA approach, we assess the beneficial environmental impacts of using fly ash to offset virgin cement inputs in a concrete beam with a compressive strength of 4 KSI (4,000 psi) and a lifespan of 75 years. It is important to note that this concrete product was selected to represent use of fly ash in any generic concrete application; the unit impact values do not reflect any assumptions specific to the life cycle of a concrete beam in BEES.[52] Furthermore, any concrete building product data set could have been used without changing the unit impact value. For further details on the life cycle inventory data used in this analysis, refer to Appendix D.

The benefits of fly ash use are measured as the difference in environmental impacts between a baseline scenario and a beneficial use scenario. In the baseline scenario, a one cubic yard 4 KSI concrete beam is produced using 100% portland cement. In the beneficial use scenario, a one cubic yard 4 KSI concrete beam is produced using 15% coal fly ash and 85% portland cement. [53, 54] The difference in environmental impacts between the baseline and beneficial use scenarios represents the change in impacts from substituting 15% of the portland cement with fly ash in one cubic yard of 4 KSI concrete. We translate these impacts from a cubic yard concrete basis to a ton fly ash basis by dividing the impacts by the absolute quantity of fly ash in one cubic yard of the concrete product. For an illustration of this methodology, refer to Appendix D.

SimaPro Analysis of FGD Gypsum in Wallboard

We calculate the unit impacts of using FGD gypsum in place of virgin gypsum stucco in wallboard as the difference in impacts between wallboard made with 100% virgin gypsum and wallboard made with 100% FGD gypsum. We model these impacts as one ton of avoided "stucco" manufacture in SimaPro.[55] Stucco is the term used in SimaPro to describe the gypsum material used in wallboard. We selected the EcoInvent data set because it includes gypsum mining but also includes the processing of gypsum for use in wallboard (i.e., burning of gypsum and milling of stucco for use in gypsum wallboard). Thus, this dataset includes all the processes that would be avoided if an equivalent quantity of FGD gypsum were used in place of stucco in wallboard. The production of FGD gypsum from coal combustion is not modeled, as discussed in the following section on allocation of life cycle impacts to FGD gypsum. In addition, this analysis assumes that FGD gypsum dewatering occurs via holding ponds and that the environmental impacts of dewatering are negligible.[56] This analysis also does not model transport distance; we assume FGD gypsum would have the same transport distances to the construction site as virgin gypsum.[57] Thus, avoided gypsum mining and avoided processing of gypsum into stucco, as represented by the EcoInvent stucco manufacturing data set, are the only lifecycle stages modeled in SimaPro. Appendix D provides more information on the FGD gypsum analysis.

Allocation of Life Cycle Impacts to CCPs

As EPA programs evolve to emphasize both beneficial use of industrial materials and life cycle analysis approaches to evaluating these programs, it is important to consider upstream impacts of the processes that create beneficial use materials, including, in this case, the impacts of the electrical power generation industry that produces CCPs for beneficial use.

The beneficial use of CCPs has positive environmental and energy impacts relative to landfilling and virgin material production. Consideration of the negative upstream impacts of electricity production through "allocation" does not actually reduce the beneficial impacts of beneficial use, nor does it "create" negative impacts. Instead, it represents a quantitative way of recognizing that CCPs are associated with the generation of coal-fired power and not an environmentally “free” product.

Analysis of life cycle impacts is, in its simplest form, the calculation of all impacts associated with a single production system (e.g., the manufacture of paper, or the production of energy using coal). However, when one production system (or a set of linked production systems) makes two or more products with market value (i.e., co-products), it is accepted practice in life cycle analysis to recognize that these products are associated with environmental impacts, and to *allocate* the total life cycle production impacts across these products.[58] Several methods for allocation are possible, depending on the system(s), inputs, and the quantity and value of co-products. Simple methods include allocating impacts proportionately by total mass or by market value; more complex methods may be necessary when integrated systems use different types of inputs or produce a range of products with different features and environmental profiles.

Waste is not considered a co-product, and it is, therefore, generally unnecessary to allocate specific production impacts to materials that are destined for disposal (disposal impacts are allocated 100% to the producing industry). However, when an industrial material becomes a beneficial use material and ceases to be considered a waste, it reflects a market value. It is, therefore, a co-product, though typically a very low-value one when compared to the primary products of the industry (in this case, electricity). It is important to consider whether co-products of electricity generation (such as fly ash and FGD gypsum) that are beneficially used should have some portion of the production impacts associated with coal combustion (e.g., energy use, greenhouse gas equivalents, releases to air and water) attributed to them.[59]

In Appendix E, we provide an illustration of a potential approach for allocating the environmental and energy impacts from coal-fired power generation across electricity generation and CCPs. The analysis considers some hypothetical macro-level scenarios for coal-fired power generation, as well as macro-level flows of several key CCPs. The preliminary analysis in Appendix E is designed to assess the implications of allocating the environmental effects of power generation to both the energy product and the CCPs. Using both an economic and a mass-based approach, we find that the only small flows would be allocated to the CCPs relative to the impacts of electricity production (i.e. less than one percent in the case of mass- based allocation).

Because of the small environmental impacts allocation indicated by our preliminary analysis in Appendix E, and because of high uncertainty associated with fuel sources, prices of electricity generation, and CCP prices, we do not currently include either an economic or mass-based allocation of coal combustion impacts to fly ash or FGD gypsum into this analysis. However, to fully understand the potential impacts of beneficial use on coal combustion, and to fully characterize the benefits associated with beneficial use, it may be important to assess these impacts under various analytical scenarios as the program moves forward.

Typically, as with fly ash and FGD gypsum, the economic value of beneficial use materials is small in comparison to the value of primary products of the producing industries. However, it is conceivable that significant increases in the value of beneficial use materials could alter the economics of the producing industries. While it is unlikely that any industry would alter production to *increase* production of beneficial use materials, demand for these materials could improve the cost structure of certain industrial processes. For example, increased demand for CCPs could improve the cost structure for coal-fired power plants and improve their competitive position in energy markets.[60] As beneficial use and the economic value of various industrial materials increases, it becomes increasingly important to accurately account for the processes that produce the materials as well as the processes that use them.

Results

For both the baseline and beneficial use scenarios, BEES and SimaPro generate quantitative estimates of impacts for a suite of environmental metrics. For each metric, environmental outputs under the baseline and beneficial use scenarios represent life cycle impacts of replacing virgin materials with CCPs. Where this difference is positive, the impact is an environmental benefit of using CCPs in place of virgin materials. Where the difference is negative, use of CCPs suggests a decline in environmental quality. Exhibit 4-1 presents the results of the analyses of use of fly ash in concrete and use of FGD gypsum in wallboard.

Exhibit 4-1. Lifecycle Assessment of Potential Impacts of CCP Beneficial Use

AVOIDED IMPACTS	PER 1 TON FLY ASH AS PORTLAND CEMENT IN CONCRETE	PER 1 TON FGD GYPSUM IN WALLBOARD
ENERGY USE		
NONRENEWABLE ENERGY (MJ)[a]	4,214.18	12,568.97
RENEWABLE ENERGY (MJ)[b]	43.55	13.69
TOTAL PRIMARY ENERGY (MJ)[c]	4,259.29	12,582.66
TOTAL PRIMARY ENERGY (US$)[d]	119.26	352.31
WATER USE		
TOTAL WATER USE (L)	341.56	14,214.60
TOTAL WATER USE (US$)[e]	0.22	9.01
GREENHOUSE GAS EMISSIONS		
CO_2 (G)	636,170.21	77,754.24
METHANE (G)	539.49	175.51
AIR EMISSIONS		
CO (G)	593.45	39.06
NOX (G)	1,932.48	168.02

Exhibit 4-1. (Continued)

AVOIDED IMPACTS	PER 1 TON FLY ASH AS PORTLAND CEMENT IN CONCRETE	PER 1 TON FGD GYPSUM IN WALLBOARD
SOX (G)	1,518.21	139.14
PARTICULATES GREATER THAN PM_{10} (G)	0.00	1,194.25
PARTICULATES LESS THAN OR EQUAL TO PM_{10} (G)	0.01	520.93
PARTICULATES UNSPECIFIED (G)	1,745.25	17.11
MERCURY (G)	0.04	0.00
LEAD (G)	0.03	0.03
WATERBORNE WASTES		
SUSPENDED MATTER (G)	13.96	23.60
BIOLOGICAL OXYGEN DEMAND (G)	3.07	21.87
CHEMICAL OXYGEN DEMAND (G)	26.00	24.71
COPPER (G)	0.00	0.02
MERCURY (G)	0.00	0.00
LEAD (G)	0.00	0.01
SELENIUM (G)	0.00	0.00
NONHAZARDOUS WASTE (KG)[f]	0.00	3.12

Notes:

a. Nonrenewable energy refers to energy derived from fossil fuels such as coal, natural gas and oil.

b. Renewable energy refers to energy derived from renewable sources, but BEES does not specify what sources these include.

c. Total primary energy refers to the sum of nonrenewable and renewable energy.

d. In addition to reporting energy impacts in megajoules (MJ), we monetize impacts by multiplying model outputs in MJ by the average cost of electricity in 2006 ($0.0275/MJ), converted to 2007 dollars ($0.0280/MJ). The 2006 cost of energy is taken from the Federal Register, February 27, 2006, accessed at: http://www.npga.org/14a/pages/index.cfm?pageid=914. The cost was converted to 2007 dollars using NASA's Gross Domestic Product Deflator Inflation Calculator, accessed at: http://cost.jsc.nasa.gov/inflateGDP.html.

e. In addition to reporting water impacts in gallons, we monetize impacts by converting model outputs from liters to gallons and multiplying by the average cost per gallon of water between July 2004 and July 2005 ($0.0023/gal), converted to 2007 dollars ($0.0024/gal). The 2005 cost of water is taken from NUS Consulting Group, accessed at: https://www.energyvortex.com/files/NUS_quick_click.pdf. The cost was converted to 2007 dollars using NASA's Gross Domestic Product Deflator Inflation Calculator, accessed at: http://cost.jsc.nasa.gov/inflateGDP.html.

f. BEES reports waste as "end of life waste." In contrast, SimaPro reports "solid waste." In is not clear if these waste metrics are directly comparable as SimaPro does not specify whether "solid waste" refers to manufacturing waste, end-of-life waste, or both.

As shown in Exhibit 4-1, the results of the fly ash and FGD gypsum analyses suggest many positive environmental impacts associated with beneficial use. For most metrics, there is a significant difference between the unit impact value for fly ash and FGD gypsum. The difference in unit impact values reflects different avoided processes when fly ash is used to offset portland cement versus when FGD gypsum is used to offset virgin gypsum. For example, the primary driver of benefits when fly ash is used in concrete is avoided raw materials extraction and avoided portland cement production.[61] In comparison, the primary driver of benefits when FGD gypsum is used in wallboard is avoided virgin gypsum extraction and the processing of virgin gypsum into stucco. Portland cement production generates relatively high greenhouse gas emissions. Thus, the avoided CO_2 and methane emissions are greater for fly ash than for FGD gypsum in this analysis. In contrast, gypsum mining requires comparatively higher quantities of water, so the water savings are greater for FGD gypsum in this analysis than for portland cement.

In addition, the difference in unit impacts may reflect differences in the assumed system boundaries between the two analyses. It is unclear how the BEES system boundaries compare to SimaPro. Thus, the total life cycle impacts calculated in BEES could be large or small in comparison to the system boundaries in the SimaPro FGD gypsum analysis.

Limitations and Assumptions

Although the BEES analysis provides a useful example of the benefits that can be achieved through beneficial use of fly ash in concrete, it is important to recognize some of the key limitations and assumptions of the work to date:

- The BEES model may over- or underestimate the national impacts of using fly ash in concrete construction projects because site-specific environmental conditions and proximity to sources of fly ash may affect the resulting benefits and influence the net effect of choosing fly ash over portland cement.
- BEES assumes round-trip distances for the transport of concrete raw materials to the ready-mix plant of 60 miles for portland cement and fly ash and 50 miles for aggregate. The user cannot adjust these transport distances. This analysis also assumes the minimum possible transport distances for the finished concrete products to the construction site. This transport distance for ready-mix concrete for a pavement application is 50 miles.
- BEES environmental results are reported in physical quantities (e.g., MJ energy, liters water, g CO, g NO, g Hg, etc.), not in monetized terms.
- In BEES, the calculation of each environmental impact is not fully transparent. BEES does disaggregate the total life cycle impact value for each environmental metric (e.g., energy use, CO_2 emissions, etc.) by lifecycle stage and by product component, but it is not possible to see exactly how each impact is derived. This limits the user's ability to compare the results of the BEES model with those of others models, such as SimaPro.
- The FGD gypsum analysis is based on a Swiss life cycle inventory data. While we substituted Swiss electricity data with the average U.S. energy mix, it is unclear whether the average U.S. energy mix is an accurate representation of the electricity mix used in wallboard manufacturing. Given the recent trend in new wallboard facilities being constructed adjacent to coal-fired powered plants, it is possible that

these facilities use primarily coal-based electricity. With the exception of energy mix, it is unlikely that any other differences between European and U.S. gypsum extraction and stucco processing would result in meaningful differences in environmental impacts.
- The FGD gypsum analysis assumes that dewatering of FGD gypsum is accomplished through evaporation in holding ponds. To the extent that the predominant practice is to use mechanical dewatering processes, the analysis should be modified to reflect this. The assumption of dewatering via holding ponds likely overstates the energy and energy-related emissions impacts in this analysis, since the impacts of dewatering, which are subtracted from the avoided gypsum processing impacts, would be greater for mechanical dewatering than for holding pond evaporation.

5. Estimating Program Level Impacts

This chapter provides an overview of an initial, life-cycle based approach to evaluating program level impacts associated with the RCC effort to increase beneficial use of CCPs. The chapter first outlines two critical steps necessary for a complete evaluation of specific program impacts:

- Development of defensible beneficial use scenarios that reflect likely market trends, policy efforts, and key limitations; and
- Implementation of a well-supported attribution protocol for assigning beneficial use impacts to specific EPA programs.

This discussion is followed by a preliminary analysis of the total impacts associated with current (baseline) beneficial use patterns, based on an extrapolation of the life cycle analysis impacts identified in Chapter 4. The purpose of this chapter is to present an initial estimate of the measurable impacts associated with current levels of beneficial use of CCPs, and to outline the steps necessary to provide a refined, program-specific analysis of EPA's efforts through the RCC to increase beneficial use of CCPs.

Development of Defensible Beneficial Use Scenarios

Life cycle inventories (LCI) and LCA provide comprehensive information on the impacts associated with given quantities of materials used in specified systems. The impacts measured by LCA models are typically linear; as the quantity of CCPs used in a particular application (e.g., concrete) is increased, the environmental impacts increase proportionately.

While LCA can provide insights into the potential magnitude of program benefits, in some cases existing market limitations and trends suggest that a linear extrapolation of current practices would be unrealistic. An effective assessment of true program impacts requires the development of defensible market scenarios that accurately identify the extent to which different beneficial uses are likely to increase, given the realities of the existing and emerging markets for beneficial use and the structure of RCC programs.

Current Market Dynamics: Factors Affecting Beneficial Use

Several market factors can limit the increased beneficial use of CCPs in various products. In some cases programs can be designed to address these factors effectively. Exhibit 5-1 outlines several of these factors and presents hypothetical actions that might address them. It is important to note that the actions described below are intended only to illustrate possible conditions for increasing the beneficial use of CCPs; they do not represent specific policy recommendations or existing program priorities.

Exhibit 5-1. Limiting Factors to Increased Beneficial Use of CCPs

FACTOR TYPE	FACTORS AFFECTING INCREASED BENEFICIAL USE	HYPOTHETICAL ACTIONS TO INCREASE BENEFICIAL USE
Economic	Transportation costs generally limit the shipment of CCPs to within about a 50 to 150 mile radius of power plants. In some cases, however, the cost of transport to the end user may be prohibitively expensive.	Implementation of strategic actions to create incentives to increase beneficial use by shifting the economic drivers (i.e., cost of materials) in favor of CCPs. Potential incentives could include tax credits for the use of CCPs, increased CCP landfill disposal tipping fees, or streamlining the permitting process for facilities that use CCPs near coal combustion plants (e.g., FGD gypsum plants).
	In some parts of the country and for certain use applications, the cost of virgin materials may be cheaper than CCPs.	
	Inexpensive landfill disposal can limit incentive to sell rather than dispose of CCPs.	
Institutional	National standards organizations have promulgated specifications that limit or disallow the use of CCPs in some construction applications because of quality and performance concerns and perceptions.	State DOTs rely on consensus standards for guidance and generally accept the use of fly ash in concrete. DOT projects can be used to demonstrate the performance of CCPs in geotechnical applications.
	The implementation of the U.S. Clean Air Mercury Rule (CAMR) may result in altering the chemical properties of fly ash, rendering it unmarketable for beneficial use.	Establishment of a research and development infrastructure to address the technical limiting factors to CCP use.
	Similar impacts may also occur for fly ash containing higher levels of unburned carbon or other components resulting from installation of low-NOx burners at coal-based power plants.	Provide technical and/or economic assistance to utilities using low-NOx burners to identify and implement cost-effective process modifications or new equipment to reduce the carbon content of fly ash.
Technical	Lack of consistency and quality in the production of fly ash has resulted in limited use in the high-value ready-mix concrete market. The priority at a coal-fired power plant will always be on	Taking into account the power plant's priority of generating electricity, the program could facilitate formal training programs to teach plant operators

Exhibit 5-1. (Continued)

FACTOR TYPE	FACTORS AFFECTING INCREASED BENEFICIAL USE	HYPOTHETICAL ACTIONS TO INCREASE BENEFICIAL USE
	producing electriity, not ash. A change in the combustion process, such as the type of coal burned, results in a change in ash quality, making it difficult to produce a consistent product.	about the co-product value of producing consistent-quality fly ash.
Educational	While quality and consistency of fly ash are legitimate concerns of end-users, in some cases, negative perceptions toward CCP use are unwarranted. Negative perceptions can often be attributed to a single experience using CCPs in a project that failed, even if CCPs were not the cause of the failure. For example, at one time, the Austin, TX concrete market almost turned to an all-cement market because of one misuse resulting from a lack of education about the material.	Dissemination of objective, scientific material to educate potential end users. (EPA is currently addressing this through C^2P^2 and other activities).

Sources:

1. U.S. Department of Energy, National Energy Technology Laboratory, "General Summary of State Regulations," accessed at: http://www.netl.doe.gov/E&WR/cub/states/select_state.html.
2. Energy and Environmental Research Center, "Barriers to the Increased Utilization of Coal Combustion/Desulfurization By-Products by Government and Commercial Sectors--Update 1998," EERC Topical Report DE-FC21-93MC-30097--79, July 1999.
3. American Coal Ash Association, "Frequently Asked Questions," accessed at: http://www.acaa-usa.org/FAQ.htm.
4. Schwartz, Karen D. "The Outlook for CCPs," *Electric Perspectives*, July/August 2003.
5. Energy & Environmental Research Center, University of North Dakota, "Review of Florida Regulations, Standards, and Practices Related to the Use of Coal Combustion Products: Final Report," April 2006, accessed at: http://www.undeerc.org/carrc/Assets/TBFLStateReviewFinal.pdf.
6. Energy & Environmental Research Center, University of North Dakota, "Review of Texas Regulations, Standards, and Practices Related to the Use of Coal Combustion Products: Final Report," January 2005, accessed at: http://www.undeerc.org/carrc/Assets/TXStateReviewFinalReport.pdf.

Exhibit 5-1 outlines a number of economic and non-economic factors that may limit the increased beneficial use of CCPs. The economic factors primarily relate to transportation costs and the price of virgin materials; a critical limitation of the market for CCPs may be the regional nature of the coal-burning industry and the extent to which CCPs can find viable markets competing with virgin materials within an economically viable distance.

Several of the non-economic issues presented in Exhibit 5-1 may also limit the expansion of CCP beneficial use, if, for example, new uses and new markets require changes in state policies governing CCP use. In response, targeted efforts among states to harmonize policies regarding beneficial use of CCPs could result in expansion of certain uses of CCPs.

To effectively describe impacts associated with RCC programs and goals, it is important to develop scenarios that correctly identify the limits of existing markets, and calculate the quantitative impact on the use of CCPs in different applications. In addition, scenarios should incorporate RCC priorities, in order to better predict which uses and markets are likely to expand. At this time, data on market limitations and on program priorities and goals are not refined enough to inform a detailed program analysis. However, spatial information about coal-fired power plants and new data on emerging markets may be sufficient to support an analysis in the near future. The result could be a set of scenarios that pinpoints specific uses for CCPs that are likely to grow and notes regional and national limits for certain applications.

Structural Changes to the Market

In addition to changes related to market trends and program activities, beneficial use scenarios must, in some cases, reflect significant changes to the market, such as large-scale technology shifts that might affect demand for or production of CCPs. In addition, for materials frequently used in construction, unexpected events such as large storms or terrorist attacks may result in sudden, regional changes in demand if, for example, large quantities of materials are needed for reconstruction or if coal-fired power plant operations change significantly as a result of a regional event.

Since these events are by definition unpredictable, it is important to identify methods for analyzing impacts if and when they occur. In particular, analyses that clearly identify the current regional market conditions (e.g., oversupply, strong demand) would provide a useful starting point for analysis of unexpected market shifts.

Attribution of Impacts to EPA Programs

The factors affecting increased beneficial use also link to the issue of attributing changes in beneficial use markets or behavior to specific EPA initiatives. The issue of attribution is complex, and in many cases the data necessary to support a clear attribution of impacts are not available. Particularly in the case of voluntary programs, it is often difficult to attribute changes in behavior (or a proportion of the change in behavior) to specific EPA activities. For example, changes in recycling or source reduction may be due to outside forces (i.e., market dynamics), multiple government programs, or a combination of both.

One starting point in addressing the attribution of benefits to EPA activities is an examination of existing information and methods describing the performance of target EPA programs and overall trends in beneficial use. Linking program activity with market trends can, on a qualitative basis, provide an indication of whether the program is having an effect. This initial scoping exercise can then be supplemented with the development of specific program scenarios that endeavor to quantify incremental beneficial use levels attributable to EPA's initiatives. In other cases, it may be necessary to start with the assumption that all costs and all benefits are related to EPA activities, and adjust that assumption as programs mature and data become available.

A full-scale, defensible approach to attribution of voluntary program impacts, however, requires a clear understanding of both the specific activities undertaken by the program and the differences between behavior of program participants and those who do not participate. This information can sometimes be obtained or identified through broadly collected data that includes both participants and non-participants (e.g., the Biennial Report or the Toxics Release Inventory). In other cases, behavioral research can help predict effective response rates to different types of programs. Exhibit 5-2 outlines the process for a full- scale assessment of attribution.

Exhibit 5-2. Outline to Attribute Voluntary Program Impacts

STEP 1: ASSESS THE MARKET FAILURE BEING ADDRESSED

- Identify and describe specific market failure of interest
  Example: material with market value being disposed as waste
- Evaluate size of market failure: evaluation of total quantity affected, current recovery and management, potential *economically feasible* recovery
- Identify key behavior changes necessary to address market failure
- Identify programs in place to address market failure, including Federal, State, local, and private efforts.

STEP 2: DESCRIBE IN DETAIL EACH EPA PROGRAM WORKING ON THE "ISSUE"

- Summarize program goals, structure, policy leverage points using program evaluation methodologies.
- Identify, for each relevant program, current and intended participants, key resources available, actions taken by participants, timeline for behavioral change, and link between activities and behavioral change.

STEP 3: IDENTIFY AVAILABLE DATA ON EPA AND OTHER EFFORTS

- Quantitative estimates of recent trends in target behavior, specific estimates of recent changes in behavior among EPA program members and non-members, and research on response rates for similar programs, strategies.

STEP 4: ATTRIBUTION OF IMPACTS TO EPA PROGRAM(S)

- Refine analysis of data collected in Step 3 to identify: changes in behavior among EPA program participants and among non-members, expected leverage of EPA activities across federal, state, private programs (e.g., by expansion of recycling efforts from pilot programs or harmonization of state regulations as a result of EPA information development), and expected leverage of EPA activities over time.
- Identify and correct for independent, confounding market changes that may affect the issue, such as changes in virgin raw materials prices due to sudden shortage.

The result of this approach should yield a quantitative estimate of the total extent of changes that can be attributed to EPA. Where implementation and/or tracking data are not available, approaches can potentially include theoretical estimates reflecting literature on response rates for voluntary activities. As necessary, this effort can also provide information to effectively allocate total EPA impacts across multiple programs in cases where more than one program is focused on addressing the same market failure. In areas where the attribution of outcomes is not possible due to data and methodological limitations, program structure and purpose can be revisited with the intent of developing metrics (e.g., for PART analysis) that are meaningful in measuring change without attempting to achieve a simplistic success metric of "outcome/resources."

Exhibit 5-3. Extrapolated Impacts of the Beneficial Use of CCPs

AVOIDED IMPACTS	FLY ASH IN CONCRETE EXTRAPOLATED TO RCC GOAL (18.6 MILLION TONS)[a]	FLY ASH IN CONCRETE EXTRAPOLATED TO CURRENT USE (15.0 MILLION TONS)[b]	FGD GYPSUM IN WALLBAORD EXTAPOLATED TO CURRENT USE (8.2 MILLION TONS)[c]	PARTIAL SUM OF CURRENT USE BENEFICIAL IMPACTS[d]
ENERGY USE				
NONRENEWABLE ENERGY (MJ)[e]	78.4 billion	63.2 billion	102.8 billion	166.0 billion
RENEWABLE ENERGY (MJ)[f]	810.0 million	652.8 million	111.9 million	764.7 million
TOTAL PRIMARY ENERGY (MJ)[g]	79.2 billion	63.8 billion	102.9 billion	166.7 billion
TOTAL PRIMARY ENERGY (BTU)	75 trillion	60 trillion	98 trillion	158 trillion
TOTAL PRIMARY ENERGY (US$)[h]	$2.2 billion	$1.8 billion	$2.9 billion	$4.7 billion
WATER USE				
TOTAL WATER USE (LITERS)	6.3 billion	5.2 billion	116.2 billion	121.4 billion
TOTAL WATER USE (US$)[i]	$4.0 million	$3.2 million	$73.7 million	$77.9 million
GREENHOUSE GAS EMISSIONS				
CO_2 (G)	11.8 trillion	9.5 trillion	0.6 trillion	10.2 trillion
METHANE (G)	10.0 billion	8.1 billion	1.4 billion	9.5 billion
TONS CO_2 EQUIVALENT[j]	13.2 million	10.6 million	0.7 million	11.5 million
METRIC TONS CARBON EQUIVA-LENT (MTCE)[k]	3.6 million	2.9 million	0.2 million	3.1 million
AIR EMISSIONS				
CO (G)	11.0 billion	8.9 billion	0.3 billion	9.2 billion
NOx (G)	35.9 billion	29.0 billion	1.4 billion	30.3 billion
SOx (G)	28.2 billion	22.8 billion	1.1 billion	23.9 billion
PARTICULATES GREATER THAN PM_{10} (G)	0	0	9.7 billion	9.7 billion
PARTICULATES LESS THAN OR EQUAL TO PM_{10} (G)	0.2 million	.02 million	4.3 million	4.3 million
PARTICULATES UNSPECIFIED (G)	32.5 billion	26.1 billion	0.1 billion	26.3 billion
MERCURY (G)	714,000	576,000	8,000	584,000
LEAD (G)	523,000	421,000	235,000	656,000

Exhibit 5-3. (Continued)

AVOIDED IMPACTS	FLY ASH IN CONCRETE EXTRAPOLATED TO RCC GOAL (18.6 MILLION TONS)[a]	FLY ASH IN CONCRETE EXTRAPOLATED TO CURRENT USE (15.0 MILLION TONS)[b]	FGD GYPSUM IN WALLBAORD EXTAPOLATED TO CURRENT USE (8.2 MILLION TONS)[c]	PARTIAL SUM OF CURRENT USE BENEFICIAL IMPACTS[d]
WATERBORNE WASTES				
SUSPENDED MATTER (G)	259.6 million	209.2 million	193.0 million	402.2 million
BIOLOGICAL OXYGEN DEMAND (G)	57.1 million	46.1 million	178.8 million	224.9 million
CHEMICAL OXYGEN DEMAND (G)	483.6 million	389.7 million	202.1 million	591.8 million
COPPER (G)	0	0	194,000	194,000
MERCURY (G)	1	0	3,000	3,000
LEAD (G)	0	0	65,000	65,000
SELENIUM (G)	3	2	2,000	2,000
NON-HAZARDOUS WASTE (KG)[k]	0	0	25.4 million	25.4 million

Notes:

a. We extrapolate the incremental impacts (i.e., impacts associated with use of 1 ton fly ash) to estimate impacts of attaining the RCC goal for the use of fly ash in concrete (18.6 million tons by 2011). To extrapolate, we multiply each of the incremental impacts calculated by the BEES model by 18.6 million.

b. We extrapolate the incremental impacts (i.e., impacts associated with use of 1 ton fly ash) to estimate the impacts of current beneficial use of fly ash in concrete (15.0 million tons). The current quantity of fly ash that is beneficially used as a substitute for finished portland cement in concrete is reported by ACAA's 2005 CCP Survey. We multiply each of the incremental impacts calculated by BEES by 15.0 million tons to extrapolate these impacts to reflect current use.

c. We extrapolate the incremental impacts (i.e., impacts associated with use of 1 ton FGD gypsum) to estimate the impacts of current beneficial use of FGD gypsum in wallboard (8.2 million tons). The current quantity of FGD gypsum that is beneficially used as a substitute for finished portland cement in concrete is reported by ACAA's 2005 CCP Survey. We multiply each of the incremental impacts calculated by SimaPro by 8.2 million to extrapolate these impacts to reflect current use.

d. Calculated as the sum of the fly ash and FGD gypsum current use extrapolations.

e. Nonrenewable energy refers to energy derived from fossil fuels such as coal, natural gas and oil.

f. Renewable energy refers to energy derived from renewable sources, but BEES does not specify what sources these include.

g.

h. In addition to reporting energy impacts in megajoules (MJ), we monetize impacts by multiplying model outputs in MJ by the average cost of electricity in 2006 ($0.0275/MJ), converted to 2007 dollars ($0.0280/MJ). The 2006 cost of energy is taken from the Federal Register, February 27, 2006, accessed at: http://www.npga.org/14a/pages/index.cfm?pageid=914. The cost was converted to 2007 dollars using NASA's Gross Domestic Product Deflator Inflation Calculator, accessed at: http://cost.jsc.nasa.gov/inflateGDP.html.

i. In addition to reporting water impacts in gallons, we monetize impacts by converting model outputs from liters to gallons and multiplying by the average cost per gallon of water between July 2004 and July 2005 ($0.0023/gal), converted to 2007 dollars ($0.0024/gal). The 2005 cost of water is taken from NUS Consulting Group, accessed at:

https://www.energyvortex.com/files/NUS_quick_click.pdf. The cost was converted to 2007 dollars using NASA's Gross Domestic Product Deflator Inflation Calculator, accessed at: http://cost.jsc.nasa.gov/inflateGDP.html.

j. Greenhouse gas emissions have been converted to tons of CO_2 equivalent using U.S. Climate Technology Cooperation Gateway's Greenhouse Gas Equivalencies Calculator accessed at: http://www.usctcgateway.net/tool/. This calculation only includes CO_2 and methane.

k. Impacts in MTCE are calculated by dividing the impacts in MTCO2E by 44/12 (the ratio of the molecular weight of carbon dioxide to carbon). U.S. EPA, "A Climate Change Glossary," accessed at: http://www.globalwarming.org/node/91.

l. BEES reports waste as "end of life waste." In contrast, SimaPro reports "solid waste." In is not clear if these waste metrics are directly comparable as SimaPro does not specify whether "solid waste" refers to manufacturing waste, end-of-life waste, or both.

In the absence of specific information on behavior changes among participants and non-participants in RCC beneficial use activities, we focus below on total impacts associated with CCP beneficial use. This forms an upper bound estimate of the impact of EPA programs related to these materials, but may understate total beneficial impacts because not all materials are considered.

Beneficial Impacts Associated with Current Use of CCPs

In the absence of defensible beneficial use scenarios for CCPs or a well-supported allocation protocol for assigning beneficial use impacts to specific EPA programs, we present a preliminary analysis of the total impacts associated with current (baseline) beneficial use patterns. While these impacts do not strictly reflect RCC program achievements, they represent the best available information on the environmental benefits of beneficially using CCPs, and reflect the impacts of all EPA, state, and industry efforts to increase CCP use to its 2005 level.

We calculate the beneficial impacts of current beneficial use of CCPs by extrapolating the life cycle analysis impacts identified in Chapter 4 to the current quantity of CCPs beneficially used in each application as presented in ACAA's 2005 CCP Survey. We also calculate the beneficial impacts of achieving the RCC goal for beneficial use of fly ash (i.e., use of 18.6 million tons of fly ash in concrete by 2011). We are unable to similarly calculate RCC program achievements for increased use of FGD gypsum as a program goal has not been developed for FGD gypsum. Exhibit 5-3 presents the impacts of the beneficial use of fly ash and FGD gypsum extrapolated both to current use quantities and the RCC goal for use of fly ash in concrete.

The results show that current beneficial use of fly ash in concrete and FGD gypsum in wallboard results in positive environmental impacts. The most significant impacts include energy savings and water use reductions. Energy savings associated with the use of fly ash and FGD gypsum totals approximately 167 billion megajoules of energy (or approximately $4.7 billion in 2007 energy prices). Based on the average monthly consumption of residential electricity customers, this is enough energy to power over 4 million homes for an entire year. Avoided water use totals approximately 121 billion liters or approximately $76.9 million in 2007 water prices).[62] This is roughly equivalent to the annual water consumption of 61,000 Americans.[63] The extrapolated beneficial impacts also include key impacts such avoided greenhouse gas (11.5 million tons of avoided CO_2 equivalent), and avoided air emissions (30.3 million kilograms of avoided NOx, and 23.9 million kilograms of SOx). Note that the impacts presented in Exhibit 5-3 represent only a partial estimate of the total impacts of beneficially using CCPs. Beneficial use of fly ash as a substitute for finished portland cement in concrete and FGD gypsum in wallboard accounts for only 47% (23.2 million tons) of all beneficially used CCPs in 2005.[64]

Economic Distribution of CCP Beneficial Use Impacts

In addition to an estimate of overall beneficial impacts associated with use of CCPs, we developed a screening analysis using the EIO-LCA model to provide insight into the distribution of impacts across economic sectors. We modeled the impacts associated with a hypothetical reduction of $1 million of demand from the cement manufacture and gypsum mining sectors. From the perspective of energy and air emissions, cement manufacturing leads to large impacts, and is in general the largest source of emissions across the supply chain. Reducing the amount of cement produced by beneficially reusing products can lead to large supply chain-wide reductions of emissions. Comparatively, the impact of the substitution of FGD gypsum for virgin gypsum in wallboard manufacturing is less clear, as the EIO-LCA model was not able to adequately represent the wallboard sector.

These results and others produced by the model do not affect the total estimate of beneficial impacts associated with changes in use of CCPs. However, they indicate the specific sectors, activities, and points in the supply chain that may be most important to consider in more detailed analyses of beneficial use scenarios. The EIO-LCA model may provide important insights into the success of policies and actions, because the model identifies the types of market changes that may result from specific changes in practice. EIO-LCA may, therefore, clarify the positive and negative impacts of specific, targeted programs and actions on different economic sectors. Appendix C provides a detailed discussion of the analysis.

Conclusions and Next Steps

This chapter provides a preliminary assessment of the baseline impacts associated with the beneficial use of fly ash and FGD gypsum in 2005. The analysis uses the life cycle-based BEES and SimaPro models, coupled with simple monetized estimates of energy and water savings, to estimate the impacts of replacing portland cement with fly ash in concrete and virgin gypsum with FGD gypsum in wallboard. The most significant impacts include energy savings and water use reductions. Energy savings associated with the use of fly ash and FGD gypsum totals approximately 167 billion megajoules of energy (or approximately $4.7 billion in 2007 energy prices). Based on the average monthly consumption of residential electricity customers, this is enough energy to power over 4 million homes for an entire year. Avoided water use totals approximately 121 billion liters or approximately $76.9 million in 2007 water prices).[65] This is roughly equivalent to the annual water consumption of 61,000 Americans.[66] The extrapolated beneficial impacts also include key impacts such avoided greenhouse gas (11.5 million tons of avoided CO_2 equivalent), and avoided air emissions (30.3 million kilograms of avoided NOx, and 23.9 million kilograms of SOx).

This chapter also presents a distributional screening analysis using the EIO-LCA model that indicates significant avoided environmental impacts from reductions in the demand for cement or virgin gypsum that are distributed across several economic sectors. From the perspective of energy and air emissions, cement manufacturing leads to large impacts, and is in general the largest source of emissions across the supply chain. Reducing the amount of cement produced by beneficially reusing products can lead to large supply chain-wide reductions of emissions. Comparatively, the impact of the substitution of FGD gypsum for virgin gypsum in wallboard manufacturing is less clear.

The preliminary results of this initial analysis suggest that a more detailed evaluation of the beneficial use of CCPs could build on these results to assist the Agency in a more specific evaluation of the achievements of the RCC program. A more detailed analysis would require:

- The development of realistic and effective beneficial use scenarios that incorporate more detailed descriptions of markets, beneficial uses, and policies. Realistic scenarios should reflect key market dynamics and limits such as distance to markets and virgin material prices, and be able to assess the impacts of these dynamics on the growth potential for specific beneficial uses. For example, the limiting transportation distance for the beneficial use of CCPs in road construction may be far less then that of gypsum wallboard.
- The development of a methodology to attribute beneficial use impacts to specific EPA/RCC efforts and programs. A phased approach may be appropriate. Such an approach could initially employ the simple operating assumption that all impacts result from Agency actions. This assumption could then be refined to reflect specific Agency strategies, policies, and other efforts, and link these, where possible, to specific changes in beneficial use practices and markets.
- The expansion of the assessment to include additional CCPs and beneficial use applications. This analysis only examines the beneficial impacts of substituting fly ash for finished portland cement in concrete and substituting FGD gypsum for virgin gypsum in wallboard manufacturing. These two processes represent less than 50% of the total beneficial use of CCPs. Additional high volume applications that may be analyzed include: the use of fly ash as a raw feed in cement clinker; the use of boiler slag as blasting grit; and the use of various CCPs in structural fill and waste stabilization. In addition, the beneficial impacts of lower volume applications may be examined in order to identify those that may have potentially high incremental impacts.

Appendix A. Key Beneficial Use Applications for CCPs

Exhibit A-1. Key Beneficial Use Applications for CCPs (2005)

BENEFICIAL USE APPLICATION AND INDUSTRY	FLY ASH	BOTTOM ASH	FLUE GAS DESULFUR-IZATION GYPSUM	FLUE GAS DESULFUR-IZATION OTHER WET MATERIAL	FLUE GAS DESULFUR-IZATION DRY MATERIAL	BOILER SLAG	PRODUCT SUBSTITUTES
Concrete	14,989,958	1,020,659	328,752	0	13,965	0	Cement, Silica fume, Furnace slag
Cement additive	2,834,476	939,667	397,743	782	0	42,566	Clay, Soil, Shale, Gypsum
Flowable fill	88,549	0	0	0	9,673	0	Soil, Sand, Gravel, Cement
Structural fill	5,710,749	2,321,140	0	0	2,666	175,144	Sand, Gravel, Soil, Aggregate
Road base	205,032	1,056,660	0	0	0	300	Cement, Lime, Aggregate
Soil stabilizer	715,996	205,322	0	0	1,535	0	Cement, Lime, Aggregate
Mineral filler in asphalt	62,546	21,583	0	0	0	56,709	Sand
Snow and ice control	591	531,549	0	0	0	15,401	Sand
Blasting grit	0	89,109	0	0	0	1,544,298	Sand
Mine reclamation	626,428	46,604	0	245,471	112,100	31,540	Soil
Wallboard	0	0	8,178,079	0	0	0	Natural gypsum
Waste stabilization	2,657,046	42,353	0	0	0	0	Cement, Lime, Cement kiln dust
Agricultural soil amendment	23,856	7,670	361,644	3,312	19,259	0	Liming agents
Manufactured aggregate	180,275	692,501	0	0	0	0	Sand, gravel, aggregate
Miscellaneous/other	1,022,952	567,155	2,147	436,619	0	24,851	
CCP Category Use Totals	29,118,454	7,541,972	9,268,365	689,184	159,198	1,890,809	
CCP Utilization Rate	41%	43%	77%	4%	11%	97%	

Sources: 1. American Coal Ash Association. “2005 Coal Combustion Product (CCP) Production and Use Survey,” accessed at: http://www.acaausa.org/PDF/2005_CCP_Production_and_Use_Figures_Released_by_ACAA.pdf.

2. Western Region Ash Group, “Applications and Competing Materials, Coal Combustion Byproducts,” accessed at: http://www.wrashg.org/compmat.htm.

Note: Results from the 2006 CCP Production and Use Survey conducted by the ACAA indicate a total fly ash utilization rate of 44.78 percent, up from 40.95 percent (rounded to 41% above) reported for 2005. This reflects an ongoing upward trend in the CCP utilization rate over the past decade. The 2006 results were received too late for incorporation into the benefits analysis.

APPENDIX B. USE OF LIFE CYCLE ANALYSIS IN AN EVALUATION OF ECONOMIC BENEFITS

Life cycle assessment (LCA) inventory analyses of the type presented in this chapter deliver incremental changes in physical inputs, outputs, and energy arising from management or regulatory changes to an industrial production process. This discussion addresses the following issues: how does LCA relate to economic analysis of benefits, and how are economic impacts derived from changes in "physical inventory," such as energy use and alternative waste streams?

Relationship between LCA and Economic Analysis of Benefits

LCA is a performance assessment tool – a method to depict physical outcomes that can be used to assess impacts and measure progress over time. And because LCA is a systems approach to assessment, it offers significant improvement over less comprehensive techniques. However, in economic terms, LCA is only one component of a true analysis of benefits – albeit a central component.

Consider the architecture of an economic benefit assessment. At the "front end" lies a set of economic drivers that determine technologies and practices employed by industry. These drivers include raw material prices, other input costs (including transport), competitive factors, regulation, technology, and taxes. EPA programs such as RCC work to facilitate changes in the economic drivers of waste generation, handling, and disposal (e.g., a change in tipping fees, tighter permit requirements on landfills, benefits to participation in beneficial use programs, etc.). Changes in these economic drivers can be expected to lead to changes in the physical system of production. In other words, the physical system and its outputs are properly thought of as the end product of a set of economic incentives (prices) and constraints (technology).

LCA depicts production as a *system* of sometimes reinforcing, sometimes counteracting physical outcomes. In particular, it allows the analyst to predict the *incremental* physical consequences of a change in disposal practices, technology, or price incentives. Any change in the physical system leads to a corresponding cascade of system changes – as inputs are substituted, exposure pathways are changed, and technology adapts. LCA produces the net result of these various changes and, thus, the true, incremental effect on physical outputs.[67]

Deriving an incremental physical effect from a complex system is difficult enough. As the agency seeks performance measures to satisfy its GPRA and PART requirements, LCA is a natural starting point. It demands systems thinking, properly views outcomes as changes to baseline conditions, identifies tradeoffs, and yields concrete, measurable metrics. LCA can tell us *who* and *what* will be affected by changes in industrial practice, and even *where* changes are likely to occur.

However, while LCA is a fundamental building block of benefit assessment, LCA does not itself yield the social benefits and costs of industrial change. To do that, we must apply economic valuation techniques to the physical outputs of LCA analysis.

Economic Assessment is desirable because we don't really care about physical outcomes, we care about what those outcomes imply for human well-being. Another way of putting this is, how do we compare the "apples" of one change to the "oranges" of another?[68] Also, how do we compare a given "small" physical gain in one waste to a "large" reduction in another. In physical terms, we might be tempted to say that the large gain outweighs the small loss. Of course, small physical changes can have large health and environmental consequences with large economic ramifications (think of the effect of radiation or toxics on health).

To understand how energy and raw materials use and emissions of different kinds affect wellbeing we must make a set of additional "translations." A physical change in lead concentrations leads, via ecological and epidemiological processes, to changes in human exposure. Changes in exposure lead to morbidity and mortality effects. Morbidity and mortality effects have social benefits and costs.[69] Those benefits and costs are the ultimate goal of our analysis. In another example, the effect of water consumption on well-being depends upon the location, timing, and quality of the water that is consumed. The value of that water depends on how it would otherwise be used – for human consumption, industrial uses, habitat support, irrigation, etc. LCA tells us little, if anything, about these relationships. Thus, LCA may tell us relatively little about the actual welfare effects of changes in industrial process.

Unfortunately, the translation of physical changes into economic outcomes is costly, difficult, and often controversial when applied to human health or environmental outcomes. As the report notes earlier, "with the exception of water and energy savings for which current price data are available, we do not calculate these benefits in dollar terms because monetizing involves complex valuation procedures." Putting economic value on even a small set of physical impacts can be a significant and expensive proposition.

Accordingly, LCA should be regarded, not only as a necessary ingredient, but also as a practical alternative to real benefit assessment. While economic benefits are the ultimate performance measure, businesses and governments routinely rely on simpler – though imperfect – proxies to facilitate management and performance assessment. As proxies, LCA outputs are a legitimate and defensible compromise.

Inventory Changes and Welfare: The Translation of LCA Outputs to Economic Impacts

There are two basic steps that must be employed to translate LCA-generated inventories into social benefits. The first is the translation of LCA inventories into "final economic goods." The second is the valuation of those final goods.

Mapping LCA inventories into final economic goods

In general, changes in LCA physical inventories will generate a set of corresponding changes in other physical conditions relevant to human well-being. Even before economic valuation occurs, these follow-on physical implications must be assessed. For instance, to value changes in mercury releases, it is important to know how increased or decreased mercury emissions interact with exposure pathways to affect body burdens and human health. An LCA inventory does not address this issue; an analysis of epidemiology and exposure is required. Similarly, hydrological analysis is required to determine how a reduction in water usage translates into water availability in different locations and at different times. Further, ecological analysis must be deployed to answer questions such as "what is the effect of greater water availability on species and habitats?" The point is that benefit assessment requires synthetic systems thinking of an order at least as great as the original LCA analysis.[70]

The goal of these biophysical and epidemiological translations is to translate LCA inventory results to outcomes with *direct* human impact – health effects or the availability of water in a particular stream at a particular time.

In the human health realm, toxic wastes or air quality burdens must be evaluated in terms of fate, transport, and deposition models. Human health models then translate depositions into human health impacts via epidemiological analysis (e.g., dose-response relationships). EPA is relatively sophisticated in its use of such models, owing to decades of experience with air quality regulation and the analysis of economic effects arising from air quality-related health assessments.

In the ecological realm, these kinds of translations are underdeveloped. The agency is aware of this – the conclusion has been drawn from several recent SAB reports.[71] The analysis of ecological benefits is clarified by drawing distinctions between ecosystem processes and functions and the "final" outcomes of those processes (denoted here as "final ecosystem goods." Ecosystem processes and functions are the biological, chemical, and physical interactions associated with ecological features such as surface water flows, habitat types, and species populations. These functions are the things described by biology, atmospheric science, hydrology, and so on.

Final ecosystem goods arise from these components and functions but are different: they are the aspects of the ecosystem that are *directly* valued by people. The benefits of nature include many forms of recreation, aesthetic enjoyment, commercial and subsistence harvests, damage avoidance, human health, and the intangible categories mentioned earlier. Final ecosystem goods are the aspects of nature used by society in order to enjoy those benefits.

Part of the above definition is particularly important: namely, that ecosystem services are "final." Final goods are the things people actually make choices about. For an angler, these end products include a particular lake or stream and perhaps a particular species population in that water body. The choices involved include which lake, what kind of fish, what kind of boat (if any) and tackle to use, and how much time spent getting to and from the site. Valuation is about choices (is one thing better than another) and choices are the only thing economists can use to establish economic value. Environmental benefit assessment places values on the things people and households make actual choices about – the "final goods" of nature. It is very important to emphasize that many other aspects of nature are valuable, but not capable of being valued in an economic sense – because they are not subject to social or individual choices.

Ecosystem production functions are the relationships that translate LCA inventory changes into final ecosystem goods. One characteristic of these production functions is particularly worthy of note: ecological production functions are dependent upon space and landscape. Location- and scale-specificity are core characteristics of modern ecology. For example, the quality of a habitat asset can be highly dependent on the quality and spatial configuration of surrounding land uses. The ability of areas to serve as migratory pathways and forage areas typically depends on landscape conditions over an area larger than habitats relied upon directly by the migratory species. The contiguity of natural land cover patches has been shown for many species to be an indicator of habitat quality and potential species resilience. Hydrological analysis is yet another field that has long recognized the importance of relationships between landscape features. The nature of surface water flows, aquifer structures, and surface-groundwater interactions are dependent upon linked physical relationships across the landscape.

For OSW to move toward measurement of ecosystem impacts arising from beneficial use, or any other change in waste management practices, the ability to translate LCA inventory changes into final ecosystem good changes requires the development of spatial ecological modeling. Space and scale are important to the valuation of final ecosystem goods, as well.

Assigning value to changes in final ecological goods

The value of an ecosystem good is typically location-dependent. The value of a car is not closely related to whether it is located in California or New Jersey. This is not the case with ecological goods. The benefits of damage mitigation, aesthetic enjoyment, and recreational and health improvements depend on where—and when—ecosystem services arise relative to complementary inputs and substitutes. Also, the ecological asset interactions that enhance or degrade service flows are highly landscape-dependent. Accordingly, it is necessary to spatially define "service areas." An unfortunate reality is that these will be different for every identified ecosystem service. Boundaries are needed to define the likely users of a service, areas in which access to a service is possible, and the area over which services might be scarce or have substitutes. This issue is well known in environmental economics (Smith and Kopp 1996). For example, a key methodological issue in any econometric recreational benefits study is the determination of the appropriate choice set facing recreators.

While market prices can be assumed to be largely constant within a single market, there is no arbitrage to ensure this condition for the implicit prices of environmental resources. Also, many ecological services are best thought of as differentiated goods with important place-based quality differences. As noted earlier, the biophysical characteristics of ecosystems are highly landscape- dependent. The same is true of ecological services' social benefits. Accordingly, willingness to pay for ecological services is best represented by a hedonic price function, not a single price.

An intermediate step: benefit indicators as an alternative to full valuation

The spatial factors that affect ecosystem goods' value create a problem for analysts. Benefit estimates from one study in one location cannot be transferred to other sites. In practical terms, this means that ecosystem valuation is expensive, time-consuming, and difficult. Problem- specific valuation will be impractical for most regulatory applications.

In this context, one alternative to full-scale valuation is the use of "benefit indicators" (Boyd 2004, Boyd and Wainger 2002). The benefits of a given ecosystem good are affected by the following: the ecosystem feature's scarcity, natural and built substitutes, complementary inputs, and the number of people in proximity to it. All of these can and should be measured spatially. Benefit indicators are map-able, countable landscape features that affect the value of a particular ecosystem good. Benefit indicators are an input to a wide variety of tradeoff analysis approaches, but do not themselves make or calculate the results of such tradeoffs. First, they can be used as ends in themselves as regulatory or planning performance measures. Second, they can be used as part of public processes designed to elicit public preferences over environmental and economic options – as in mediated modeling exercises or more informal political derivations. Benefit indicators are a potentially powerful complement to group decision processes. Third, they can be used as *inputs* to economic and econometric methods such as benefit transfer, or stated preference models. In the former, they can be used to calibrate the transfer function. In the latter case, they can be used to develop alternative choice scenarios.

APPENDIX C. ANALYSIS OF BENEFITS USING ALTERNATE LIFE CYCLE MODELS

In the main body of this chapter, we present an analysis of the life cycle benefits of substituting fly ash for finished portland cement in concrete using BEES. For comparative purposes, this appendix illustrates the impacts associated with beneficial use of fly ash in concrete that can be calculated using two additional life cycle tools: the WARM and EIO-LCA models.

Warm Model Analysis

The Waste Reduction Model (WARM) was created by EPA to help solid waste planners and organizations estimate greenhouse gas (GHG) emission reductions from several different waste management practices.[72] WARM calculates GHG emissions for baseline and alternative waste management practices, including source reduction, recycling, combustion, composting, and landfilling. The user can construct various scenarios by entering data on the amount of waste handled by material type and by management practice. WARM then automatically applies material-specific emissions factors for each management practice to calculate the GHG emissions and energy savings of each scenario. The model evaluates energy use and GHG emissions in three stages of the life cycle: (1) raw material acquisition, (2) manufacturing (fossil fuel energy emissions), and (3) waste management (carbon dioxide emissions associated with compost, nonbiogenic carbon dioxide and nitrous oxide emissions from combustion, and methane emissions from landfills). At each of these points, the study also considers transportation-related energy use and GHG emissions.

The WARM model reports avoided lifecycle GHG emissions in either metric tons CO2 equivalent ($MTCO_2E$) or metric tons CO equivalent (MTCOE), as well as energy use in BTUs. In addition, the model converts these outputs to equivalent metrics including the equivalent number of cars removed from the road in one year, the equivalent number of avoided barrels of oil burned, and the equivalent number of avoided gallons of gasoline consumed. Currently, the only CCP available for analysis using WARM is fly ash. WARM calculates GHG emissions and energy use associated with use of fly ash in concrete as an alternative to landfill disposal. We first use WARM to estimate the incremental impacts associated with beneficial use of one ton of fly ash in concrete, in comparison to disposing of that ton of fly ash in a landfill. Then, we extrapolate the results to estimate benefits associated with attainment of the 2011 RCC goal of beneficially using 18.6 million tons of fly ash in concrete.[73]

Results

Exhibit C-1 presents the results of the WARM model analysis for the beneficial use of fly ash.[74] The WARM model estimates that one ton of fly ash beneficially used in concrete results in avoidance of approximately 0.91 $MTCO_2E$ of GHG emissions and 5.29 million BTUs of energy use. Extrapolating these outcomes to the 2011 RCC goal of beneficially using 18.6 million tons of fly ash in concrete, results in savings of approximately 17 million $MTCO_2E$. According to the WARM model, this is equivalent to removing 3.7 million cars from the road. In addition, attaining the 18.6 million ton fly ash goal results in 98,394 BTUs (103.8 megajoules) of avoided energy use. This energy savings is equivalent to 17 million barrels of avoided oil consumption, 787 million gallons of avoided gasoline consumption, or a reduction in annual energy use by approximately half a million households.

Limitations and Assumptions

Although the WARM analysis provides a useful example of the energy use and GHG emissions benefits that can be achieved through the beneficial use of fly ash in concrete, it is important to recognize some of the key limitations of the work to date:

- Our analysis assumes a 20-mile transport distance from the point of collection to the landfill or concrete facility. In reality, transport distances may be greater or less than 20 miles. Adjusting transport distance would effect both GHG emissions and energy use.
- Emissions factors used in WARM reflect national averages. Our analysis may therefore over or under estimate impacts for a specific region or location. In addition, we use a national average for landfill gas recovery that may also over or understate emissions for a specific landfill.
- WARM does not specifically calculate impacts on purchased energy. Purchased energy impacts may be incorporated into the avoided energy use metric, but this is not clear.
- WARM reports some environmental impacts in physical quantities (e.g., BTUs energy, lbs NOx, etc.), not in monetized dollar effects.

EIO-LCA Analysis

The goal of this task was to do a preliminary assessment of what the baseline energy and air pollution effects are for existing cement production, such that any reduction of this demand in terms of beneficially used fly ash would lead to reduced impact.

To estimate baseline impacts of cement production, the Economic Input-Output Life Cycle Assessment (EIO-LCA) model was used. EIO-LCA was developed at Carnegie Mellon and provides the capacity to evaluate economic and environmental effects across the supply chain for any of 491 industry sectors in the U.S. economy. EIO-LCA also can represent the supply chain use of inputs and resulting environmental outputs across the supply chain by using publicly available data sources from the U.S. government. By integrating economic data on the existing flow of commerce between commodity sectors with environmental data on releases and material flows generated by each sector, it is possible to estimate the additional environmental emissions caused by an increase in production within a particular sector, accounting for the supply chain. This approach can be used to avoid some of the system boundary limitations of process LCA by drawing upon data for the entire economy. The EIO-LCA model includes a variety of such impacts for the entire US economy. For a closer look at the model, visit http://www.eiolca.net/ on the Internet.

Currently, the EIO-LCA model is in active use. Since 2000, the model has registered over 900,000 uses (or over 15,000 per month). Of identifiable access sites, educational users are most common, but there is substantial use by government agencies, non-profit organizations and companies. A surprising number of foreign users exist, suggesting that international comparisons are of considerable interest.

Exhibit C-1. Warm Results: Impacts of Beneficial Use of Fly Ash In Concrete

IMPACT	INCREMENTAL IMPACT OF USING 1 TON OF FLY ASH	TOTAL IMPACTS OF MEETING RCC GOAL (18,600,000 TONS)[a]
GHG EMISSIONS AVOIDED ($MTCO_2E$)	0.91	16.93 million
EQUIVALENT NUMBER OF PASSENGER CARS REMOVED FROM ROADWAYS	*0.20*	*3.72 million*
AVOIDED ENERGY USE (BTUs)[b]	5.29 million $135,424	98,394 billion $2,52 billion
EQUIVALENT AVOIDED OIL CONSUMPTION (BARRELS)	*0.91*	*16.93 million*
EQUIVALENT AVOIDED GASOLINE CONSUMPTION (GALLONS)	*42.33*	*787.34 million*
EQUIVALENT AVOIDED HOUSEHOLDS' ANNUAL ENERGY CONSUMPTION(HOUSEHOLDS)	*0.03*	*0.56 million*

Notes:

a. The total impacts of meeting RCC goal represent the difference between beneficially using 18.6 million tons of fly ash in concrete in comparison to disposal in a landfill.

b. In addition to reporting energy avoided energy use in BTUs, by the average retail price of electricity for all sectors in 2006 ($0.0874/KWh or $0.0256/1,000 Btu). (Source: Energy Information Administration, "Electric Power Monthly – Average Retail Price of Electricity to Ultimate Customers: Total by End-Use Sector," accessed on October 10, 2006 at: <http: //www.eia.doe.gov/cneaf/electricity/epm/table5_3. html>.)

Sources: US EPA, *Solid Waste Management and Greenhouse Gases, A Life cycle Assessment of Emissions and Sinks*, 2nd Edition, May 2002. (EPA530-R-02-006) (WARM Model)

2. US EPA, *Background Document for Life cycle Greenhouse Gas Emission Factors for Fly Ash Used as a Cement Replacement in Concrete,* November 2003. (EPA530-R-03-016)

Cement Analysis

Specifically within EIO-LCA, industry "Sector #327310: Cement Manufacturing" was selected for analysis in the model. This industry comprises establishments primarily engaged in manufacturing portland, natural, masonry, pozzolanic, and other hydraulic cements. Cement manufacturing establishments may calcine earths or mine, quarry, manufacture, or purchase lime. Examples of activities in this sector:

Exhibit C-2. Top Sectors That Support Cement Manufacturing in the US (Economic)

	SECTOR	TOTAL ECONOMIC $ MILLIONS	VALUE ADDED $ MILLIONS	DIRECT ECONOMIC %
	Total for all sectors	1.90	0.992	77.9
327310	Cement manufacturing	1.06	0.543	99.6
221100	Power generation and supply	0.083	0.052	86.0
550000	Management of companies and enterprises	0.050	0.035	65.3
221200	Natural gas distribution	0.043	0.014	89.5
211000	Oil and gas extraction	0.041	0.017	2.49
420000	Wholesale trade	0.039	0.026	50.6
484000	Truck transportation	0.033	0.016	62.4
327992	Ground or treated minerals and earths manufacturing	0.027	0.017	86.6
212310	Stone mining and quarrying	0.018	0.010	82.7
212100	Coal mining	0.017	0.008	48.8

Exhibit C-3. Air Emissions of Top 10 Sectors That Support Cement Manufacturing in The US (Sorted by SO_2)

	SECTOR	SO_2 MT	CO MT	NOX MT	VOC MT	LEAD MT	PM_{10} MT
	Total for all sectors	27.3	20.4	31.7	23.6	0.005	3.96
327310	Cement manufacturing	22.4	14.6	28.6	22.4	0.005	3.48
221100	Power generation and supply	4.44	0.219	2.01	0.019	0.000	0.094
212310	Stone mining and quarrying	0.209	0.391	0.147	0.074	0	0.043
211000	Oil and gas extraction	0.040	0.068	0.030	0.046	0	0.001
221200	Natural gas distribution	0.027	0.001	0.052	0.183	0	0.001
483000	Water transportation	0.024	0.017	0.129	0.124	0	0.012
324110	Petroleum refineries	0.018	0.010	0.004	0.014	0	0.002
213112	Support activities for oil and gas operations	0.016	0.013	0.010	0.004	0	0.002
482000	Rail transportation	0.015	0.032	0.272	0.013	0	0.006
484000	Truck transportation	0.011	3.58	0.258	0.266	0.000	0.006

- Cement (e.g., hydraulic, masonry, portland, pozzolana) manufacturing
- Cement clinker manufacturing
- Natural (i.e., calcined earth) cement manufacturing

One million dollars of demand from the cement manufacturing sector was input into EIO-LCA, resulting in the summary estimate of supply-chain wide economic impacts shown in Exhibit C-2. EIO-LCA is a linear model, thus the estimates scale in a constant fashion ($2 million of production would lead to double the results listed below). However the main point of EIO-LCA is for screening purposes, thus the specific dollar values are less relevant than the broad, economy-wide boundary which is able to show where less obvious supply chain impacts might exist. These are noted below.

As shown in the "Total Economic" column of Exhibit C-2, there are significant purchases of electricity, oil and gas, etc. across the supply chain. This is due to the recognized significant fuel and energy inputs needed to produce cement. Also visible in the top 10 economic purchases are purchases from minerals, stone, and coal.

Exhibit C-2 also summarizes which of the purchases are "direct", i.e., those made directly by the cement manufacturer. For example EIO-LCA estimates that 86% of the electricity purchases across the entire supply chain are direct. That means that only 14% of total electricity purchases of cement manufacturing in the supply chain come from all other sectors' (indirect) purchases of electricity. This would include electricity bought by oil and gas production and distribution, stone and coal mining, etc. Note that this amount of direct purchases (86 percent) is a very large amount compared to the usual electricity direct purchases that come from other sectors.

EIO-LCA also displays estimates of emissions and energy use across the supply chain, as shown in Exhibits C-3 and C-4. Exhibit C-3 summarizes emissions of conventional air pollutants, and is sorted by sulfur dioxide (SO_2) emissions. Most SO_2 comes from cement manufacturing (and about 15% from power generation). While not shown explicitly in Exhibit C-3, further use of EIO-LCA shows that about 90% of nitrogen oxides and VOC emissions from cement manufacturing come from the cement manufacturing itself, 70% of carbon monoxide from the supply chain production of cement comes from cement manufacturing, followed by truck transportation. Ninety percent of PM_{10} emissions come from cement. In short, cement manufacturing itself is a very polluting process, and avoiding emissions from its manufacture can have large social benefits.

Exhibit C-4 summarizes supply chain wide use of energy and electricity for producing cement. The cement sector consumes about 80% of total supply chain primary energy use (and almost 90% of electricity, as noted above). Other sectors consuming top but less significant amounts of energy (in the form of fuels) are truck and pipeline transportation sectors and petroleum refining.

In summary, from the perspective of energy and air emissions, cement manufacturing leads to large impacts, and is in general the largest source of emissions across the supply chain. Reducing the amount of cement produced through beneficial use of fly ash can lead to large supply chain- wide reductions of emissions.

Gypsum Analysis

Within EIO-LCA, the industry "Sector #212390: Other Nonmetallic Mineral Mining" was selected for analysis. This industry is aggregated to include many products and processes (and thus is less representative of a specific industry like wallboard manufacture than the sector representing cement manufacturing above). This U.S. industry comprises establishments primarily engaged in developing the mine site, mining and/or milling, or otherwise beneficiating (i.e., preparing) natural potassium, sodium, or boron compounds, phosphate rock, fertilizer raw materials, or nonmetallic minerals. There are many products of this industry, a few of which are summarized below:

- Borate, natural, mining and/or beneficiating
- Phosphate rock mining and/or beneficiating
- Gypsum mining and/or beneficiating
- Peat grinding

Exhibit C-4. Top 10 Sectors That Use Energy Across the Supply Chain from Cement Manufacturing (Sorted by Total Energy Use)

	SECTOR	TOTAL TJ	ELECMKWH
	Total for all sectors	68.4	1.96
327310	Cement manufacturing	55.1	1.80
221100	Power generation and supply	9.70	0.001
484000	Truck transportation	0.363	0.001
486000	Pipeline transportation	0.321	0.007
S00202	State and local government electric utilities	0.283	0
327992	Ground or treated minerals and earths manufacturing	0.238	0.015
324110	Petroleum refineries	0.203	0.004
483000	Water transportation	0.164	0.000
482000	Rail transportation	0.149	0.000
211000	Oil and gas extraction	0.144	0.015

Exhibit C-5. Top Sectors That Support Nonmetallic Mineral Mining – As A Proxy for Gypsum Manufacturing - In The US (Economic)

	SECTOR	TOTAL ECONOMIC $ MILLIONS	DIRECT ECONOMIC %
	Total for all sectors	1.98	76.3
212390	Other nonmetallic mineral mining	1.08	99.4
484000	Truck transportation	0.068	73.3
550000	Management of companies and enterprises	0.067	65.9
221100	Power generation and supply	0.055	79.2
211000	Oil and gas extraction	0.054	34.4
420000	Wholesale trade	0.042	39.5
324110	Petroleum refineries	0.035	59.0
333120	Construction machinery manufacturing	0.031	86.8
533000	Lessors of nonfinancial intangible assets	0.019	24.4
531000	Real estate	0.018	17.7

Exhibit C-6. Air Emissions of the Top 10 Sectors That Support Nonmetallic Mineral Mining In The US (Sorted by SO_2)

SECTOR	SO_2 MT	CO MT	NOX MT	VOC MT	LEAD MT	PM_{10} MT
Total for all sectors	3.47	9.41	2.85	1.72	0.000	0.421
Power generation and supply	2.94	0.145	1.33	0.013	0.000	0.062
Other nonmetallic mineral mining	0.151	0.051	0.241	0.656	0	0.169
Oil and gas extraction	0.053	0.089	0.039	0.060	0	0.002
Stone mining and quarrying	0.043	0.080	0.030	0.015	0	0.009
Petroleum refineries	0.040	0.023	0.009	0.032	0	0.004
Other basic inorganic chemical manufacturing	0.033	0.004	0.002	0.002	0	0.002
Truck transportation	0.022	7.39	0.533	0.550	0.000	0.013
Support activities for oil and gas operations	0.021	0.018	0.013	0.006	0	0.003
Iron and steel mills	0.021	0.176	0.016	0.010	0.000	0.015
Rail transportation	0.016	0.034	0.297	0.014	0	0.007

To estimate baseline effects, $ 1 Million Dollars of demand from the "Other nonmetallic mineral mining" sector was input into EIO-LCA, resulting in the summary estimate of supply-chain wide economic impacts shown in Exhibit C-5. As shown in the "Total Economic" column of Exhibit C-5, there are significant purchases of electricity, oil and gas, and construction machinery, etc. across the supply chain. This is due to the recognized significant fuel and energy inputs needed to produce nonmetallic minerals like gypsum.

Exhibit C-5 also summarizes which of the purchases are "direct," (i.e., those made directly by the nonmetallic mineral company). For example EIO-LCA estimates that 80% of the electricity purchases across the entire supply chain of nonmetallic minerals mining are direct. That means that only 20% of total electricity purchases in the supply chain come from all other sectors' (indirect) purchases of electricity, including well-known electricity intensive sectors like manufacturing. This would include electricity purchased by oil and gas production and distribution, machinery manufacturing, etc. Note that this level of direct purchases (80 percent) is very large compared to the usual electricity direct purchases that come from other sectors.

EIO-LCA also displays estimates of emissions and energy use across the supply chain, as shown in Exhibits C-6 and C-7. Exhibit C-6 summarizes emissions of conventional air pollutants, and is sorted by sulfur dioxide (SO2) emissions. Most (85 percent) of SO_2 emitted across the supply chain of nonmetallic minerals comes from power generation (less than five percent from the mining of the nonmetallic minerals, an important note). While not shown explicitly in Exhibit C6, further sorting of EIO-LCA emissions data estimates that about 50% of nitrogen oxides come from power generation, followed by emissions from truck and rail transport (less than ten percent from nonmetallic mineral mining). About 40% of VOC emissions result from nonmetallic minerals mining itself, 80% of carbon monoxide from truck transportation across the supply chain with nonmetallic mineral mining representing less than one percent. About 40% of PM10 emissions come from nonmetallic mineral mining (about 15% from power generation). In short, nonmetallic mineral mining itself is a very polluting process, and avoiding emissions from its manufacture can have large social benefits, but emissions from energy production and transportation are in some cases even more important than this sector.

Exhibit C-7 summarizes supply chain wide use of energy and electricity for producing nonmetallic minerals. The nonmetallic mineral mining sector consumes about 70% of total supply chain primary energy (and almost 80% of electricity, as noted above). Other sectors consuming high but less significant amounts of energy (in the form of fuels) are power generation, truck and pipeline transportation sectors, and petroleum refining.

Exhibit C-7. Top 10 Sectors That Use Energy Across the Supply Chain from Nonmetallic Mineral Mining (Sorted by Total Energy Use)

SECTOR	TOTAL TJ	ELEC MKWH
Total for all sectors	33.2	1.59
Other nonmetallic mineral mining	22.6	1.40
Power generation and supply	6.42	0.000
Truck transportation	0.750	0.002
Petroleum refineries	0.451	0.008
Other basic inorganic chemical manufacturing	0.370	0.033
Pipeline transportation	0.342	0.008
Iron and steel mills	0.229	0.011
State and local government electric utilities	0.225	0
Oil and gas extraction	0.189	0.019
Rail transportation	0.163	0.000

Exhibit C-8. Air Emissions of Top 10 Sectors that Support Concrete Manufacturing in The US (Sorted by SO_2)

SECTOR	SO_2 MT	CO MT	NOX MT	VOC MT	LEAD MT	PM_{10} MT
Total for all sectors	6.29	16.9	7.90	5.64	0.001	1.03
Cement manufacturing	3.74	2.43	4.78	3.74	0.000	0.582
Power generation and supply	1.61	0.080	0.728	0.007	0.000	0.034
Stone mining and quarrying	0.570	1.07	0.401	0.202	0	0.118
Water transportation	0.072	0.050	0.382	0.367	0	0.035
Other basic inorganic chemical manufacturing	0.055	0.006	0.004	0.003	0	0.003
Truck transportation	0.034	11.3	0.814	0.840	0.000	0.020
Rail transportation	0.025	0.055	0.475	0.022	0	0.011
Oil and gas extraction	0.024	0.041	0.018	0.027	0	0.000
Petroleum refineries	0.021	0.012	0.005	0.017	0	0.002
Other miscellaneous chemical product manufacturing	0.013	0.000	0.015	0.003	0	0.001

Exhibit C-9. Top 10 Sectors that Use Energy across the Supply Chain from Concrete Manufacturing (Sorted by Total Energy Use)

SECTOR	TOTAL TJ	ELEC MKWH
Total for all sectors	21.6	0.655
Cement manufacturing	9.20	0.301
Power generation and supply	3.52	0.000
Ready-mix concrete manufacturing	3.12	0.080
Truck transportation	1.15	0.004
Sand, gravel, clay, and refractory mining	0.747	0.062

Context: Concrete production and wallboard manufacturing

While the beneficial use studies focus on the substitution of waste products for virgin products, and the estimates above identify the avoided energy, cost, and emissions of these substitutions, it is important to put into context the effects of the beneficial use. In this section we briefly show EIO-LCA results for the broader picture of concrete manufacturing (where fly ash is used in place of some cement) and wallboard manufacturing (where FGD gypsum is used in place of virgin gypsum). We do this to see how important these raw materials (cement and gypsum) are in the supply chain of producing these final products.

Exhibit C-8 shows the top ten sectors that contribute to air emissions, and Exhibit C-9 the top ten sectors that consume energy, in the production of concrete (using $1 million as input into the ready-mix concrete sector). Exhibit C-8 (sorted by SO_2 emissions) demonstrates that in the manufacture of concrete, the emissions from cement manufacturing account for the majority of SO_2, NOx, VOC, and PM_{10} emissions. CO emissions are dominated by truck transportation. Similarly Exhibit C-9 shows that cement manufacture represents 40% of energy use, and almost 50% of electricity use. This implies that any reduction in the amount of cement needed has a large benefit in the life cycle emissions and energy use of concrete. Thus, fly ash substitution even at 20% substitution rates is quite beneficial.

For gypsum in wallboard manufacturing, a similar method is used, but the sector used to model wallboard (Sector #327420: Gypsum Product Manufacturing) is more aggregated than that used to model concrete. This industry comprises establishments primarily engaged in manufacturing gypsum products such as wallboard, plaster, plasterboard, molding, ornamental moldings, statuary, and architectural plaster work. Gypsum product manufacturing establishments may mine, quarry, or purchase gypsum. Examples of activities in this sector include:

- Board, gypsum, manufacturing
- Gypsum building products manufacturing
- Gypsum products (e.g., block, board, plaster, lath, rock, tile) manufacturing
- Joint compounds, gypsum based, manufacturing
- Wallboard, gypsum, manufacturing

Despite the limitations in modeling wallboard as an exclusive product, Exhibits C-10 and C-11 show the results of the supply chain emissions and energy use from the gypsum product manufacturing sector in EIO-LCA (as a proxy for gypsum wallboard manufacturing).

Exhibit C-10 summarizes the air emissions across the manufacturing supply chain for gypsum products (sorted by SO_2 emissions). Recall that gypsum was modeled by production of the nonmetallic mineral mining sector (and this is the virgin product we would be replacing). Nonmetallic mineral mining is not in the top ten emissions sources in any of the tracked conventional air emissions for gypsum product manufacturing. Exhibit C-11 shows that the nonmetallic mineral mining sector represents about five percent of total energy use of gypsum products.

As compared to the results for concrete manufacturing sector above, this wallboard example is less clear. The wallboard sector was approximated by a highly aggregated sector and is not an accurate representation of wallboard manufacturing in our attempt to model gypsum substitution. This sector seems to more generally depend on stone sectors for its inputs.

Exhibit C-10. Supply Chain Emissions from the Gypsum Product Manufacturing Sector

SECTOR	SO_2 MT	CO MT	NOX MT	VOC MT	LEAD MT	PM_{10} MT
Total for all sectors	3.44	18.1	3.94	5.23	0.000	1.02
Power generation and supply	2.22	0.109	1.00	0.010	0.000	0.047
Stone mining and quarrying	0.508	0.951	0.358	0.180	0	0.105
Cement manufacturing	0.284	0.185	0.363	0.284	0.000	0.044
Water transportation	0.071	0.049	0.374	0.360	0	0.035
Paper and paperboard mills	0.060	0.406	0.085	0.033	0	0.045
Truck transportation	0.039	12.7	0.918	0.947	0.000	0.022
Oil and gas extraction	0.036	0.061	0.027	0.041	0	0.001
Petroleum refineries	0.025	0.014	0.006	0.020	0	0.003
Rail transportation	0.019	0.041	0.354	0.017	0	0.008
Natural gas distribution	0.019	0.000	0.036	0.129	0	0.001

Exhibit C-11. Energy Use for the Gypsum Product Manufacturing Sector

SECTOR	TOTAL TJ	ELEC MKWH
Total for all sectors	24.0	0.810
Gypsum product manufacturing	9.50	0.368
Power generation and supply	4.84	0.000
Paper and paperboard mills	2.16	0.130
Truck transportation	1.29	0.004
Other nonmetallic mineral mining	1.25	0.077

APPENDIX D. DETAILS OF FLY ASH AND FGD GYPSUM LIFE CYCLE ANALYSIS METHODOLOGIES

Bees Analysis of Fly Ash In Concrete

We calculate the unit impacts of using fly ash as a substitute for finished portland cement in concrete as the difference in impacts between concrete made with 100% portland cement and concrete made with 15% fly ash and 85% portland cement. Exhibits D-1 and D-2 show the lifecycle stages modeled by BEES in the production of concrete with and without blended cement. These diagrams represent the baseline and beneficial use scenarios evaluated in the fly ash analysis.

It is important to note that in Exhibit D-2, BEES does not actually model the impacts of fly ash "production" from coal combustion (i.e. BEES does not allocate electricity production impacts to fly ash).

BEES Life Cycle Inventory Data

Exhibit D-3 presents the complete BEES lifecycle inventory data for a generic concrete beam made with and without blended cement (i.e. fly ash). The data fields in Exhibit D-3 are defined as follows:

- XPORT DIST: Transport distance of concrete beam to construction site.
- FLOW: The environmental impact being reported.
- UNIT: The unit in which the environmental flow is reported.
- TOTAL: The total impact across all life cycle stages for all three components (i.e., the sum of fields COMP1, COMP2 and COMP3).
- COMP 1: The total impact across all life cycle stages for Component 1. Component 1 is the main component, which is a 1 cubic yard concrete beam.
- COMP2: The total impact across all life cycle stages for Component 2. Component 2 refers to the first installation component associated with the concrete beam, but BEES does not provide a specific definition.
- COMP3: The total impact across all life cycle stages for Component 3. Component 3 refers to the second installation component associated with the concrete beam, but BEES does not provide a specific definition.
- RAW1: Impacts associated with raw materials extraction for Component 1.
- RAW2: Impacts associated with raw materials extraction for Component 2.
- RAW3: Impacts associated with raw materials extraction for Component 3.
- MFG1: Impacts associated with manufacturing of Component 1.
- MFG2: Impacts associated with manufacturing of Component 2.
- MFG3: Impacts associated with manufacturing of Component 3.
- XPORT 1: Impacts associated with transport of Component 1.
- XPORT2: Impacts associated with transport of Component 2.

- XPORT3: Impacts associated with transport of Component 3.
- USE 1: Impacts associated with use of the total product (all three components).
- WASTE 1: Impacts associated with disposal of the total product (all three components).

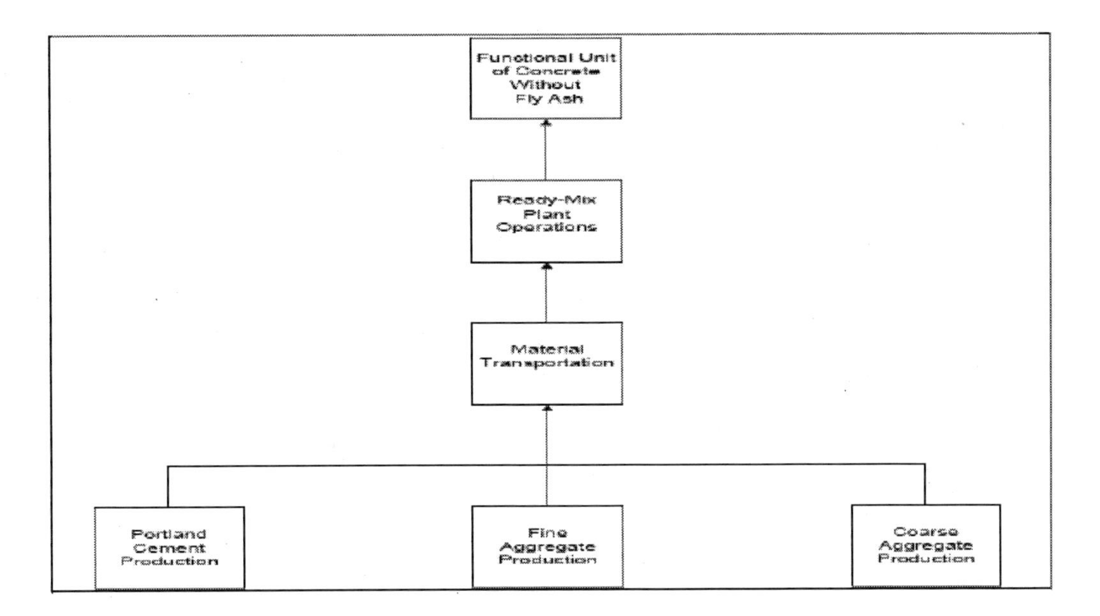

Exhibit D-1. Concrete without Blended Cement Flow-Chart (Baseline Scenario)

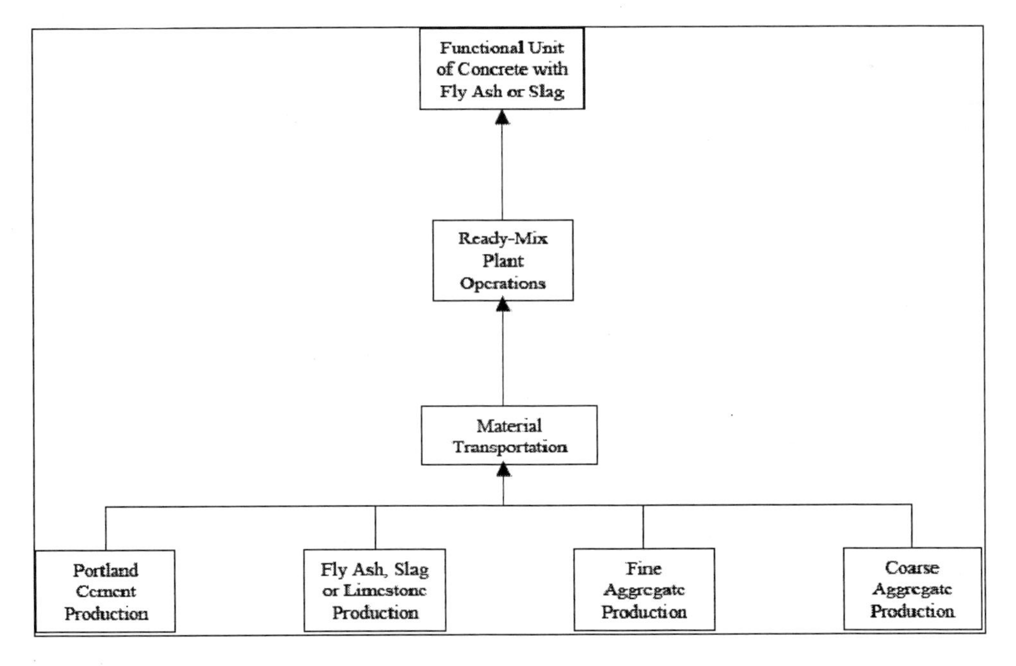

Exhibit D-2. Concrete with Blended Cement Flow-Chart (Beneficial Use Scenario)

Exhibit D-3. Bees Life Cycle Inventory Data for Concrete Beam without Blended Cement

BEES Data file B1011A: Generic Concrete Beam, 100% Portland Cement (4KSI)																	
XPORT DIST	FLOW	UNIT	TOTAL	COMP1	COMP2	COMP3	RAW1	RAW2	RAW3	MFG1	MFG2	MFG3	XPORT1	XPORT2	XPORT3	USE1	WASTE1
20	Water Used (total)	liter	1,702.10	1,055.10	570.94	4.39	1,011.14	570.02	4.25	6.05	0.00	0.07	37.91	0.92	0.08	71.67	71.67
20	Concrete Beam	Cu yd	1.00	0.00	0.00	0.00	0.00	0.00	0.00	0.00	0.00	0.00	0.00	0.00	0.00	0.00	0.00
20	Installation component 1	kg	65.77	0.00	65.77	0.00	0.00	0.00	0.00	0.00	0.00	0.00	0.00	0.00	0.00	0.00	0.00
20	Main component	kg	1,817.58	1,817.58	0.00	0.00	0.00	0.00	0.00	0.00	0.00	0.00	0.00	0.00	0.00	0.00	0.00
20	Installation component 2	kg	28.57	0.00	0.00	28.57	0.00	0.00	0.00	0.00	0.00	0.00	0.00	0.00	0.00	0.00	0.00
20	Component 4	kg	0.00	0.00	0.00	0.00	0.00	0.00	0.00	0.00	0.00	0.00	0.00	0.00	0.00	0.00	0.00
20	Component 5	kg	0.00	0.00	0.00	0.00	0.00	0.00	0.00	0.00	0.00	0.00	0.00	0.00	0.00	0.00	0.00
20	Component 6	kg	0.00	0.00	0.00	0.00	0.00	0.00	0.00	0.00	0.00	0.00	0.00	0.00	0.00	0.00	0.00
20	(a) Carbon Dioxide (CO2, fos	g	266,110.00	213,972.00	50,991.90	1,146.09	207,804.00	50,863.70	815.22	862.43	0.00	319.85	5,305.62	128.19	11.02	0.00	0.00
20	(a) Carbon Tetrafluoride (CF	g	0.00	0.00	0.00	0.00	0.00	0.00	0.00	0.00	0.00	0.00	0.00	0.00	0.00	0.00	0.00
20	(a) Lead (Pb)	g	0.43	0.01	0.42	0.00	0.01	0.42	0.00	0.00	0.00	0.00	0.00	0.00	0.00	0.00	0.00
20	(a) Mercury (Hg)	g	0.01	0.01	0.00	0.00	0.01	0.00	0.00	0.00	0.00	0.00	0.00	0.00	0.00	0.00	0.00
20	(a) Methane (CH4)	g	297.63	206.68	88.66	2.29	202.58	88.57	1.55	0.57	0.00	0.73	3.52	0.09	0.01	0.00	0.00
20	(a) Nitrogen Oxides (NOx as	g	1,299.12	1,1 71.98	118.58	8.56	1,096.00	117.07	4.87	13.60	0.00	3.56	62.38	1.51	0.13	0.00	0.00
20	(a) Nitrous Oxide (N2O)	g	7.10	6.71	0.28	0.12	5.95	0.26	0.08	0.03	0.00	0.04	0.73	0.02	0.00	0.00	0.00
20	(a) Particulates (PM 10)	g	0.01	0.01	0.00	0.00	0.01	0.00	0.00	0.00	0.00	0.00	0.00	0.00	0.00	0.00	0.00
20	(a) Sulfur Oxides (SOx as SO	g	608.93	479.47	125.58	3.88	471.71	125.41	2.64	0.71	0.00	1.23	7.06	0.17	0.01	0.00	0.00
20	(s) Aluminum (Al)	g	0.00	0.00	0.00	0.00	0.00	0.00	0.00	0.00	0.00	0.00	0.00	0.00	0.00	0.00	0.00
20	(s) Arsenic (As)	g	0.00	0.00	0.00	0.00	0.00	0.00	0.00	0.00	0.00	0.00	0.00	0.00	0.00	0.00	0.00
20	(s) Cadmium (Cd)	g	0.00	0.00	0.00	0.00	0.00	0.00	0.00	0.00	0.00	0.00	0.00	0.00	0.00	0.00	0.00
20	(s) Carbon (C)	g	0.00	0.00	0.00	0.00	0.00	0.00	0.00	0.00	0.00	0.00	0.00	0.00	0.00	0.00	0.00
20	(s) Calcium (Ca)	g	0.00	0.00	0.00	0.00	0.00	0.00	0.00	0.00	0.00	0.00	0.00	0.00	0.00	0.00	0.00
20	(s) Chromium (Cr III, Cr VI)	g	0.00	0.00	0.00	0.00	0.00	0.00	0.00	0.00	0.00	0.00	0.00	0.00	0.00	0.00	0.00
20	(s) Cobalt (Co)	g	0.00	0.00	0.00	0.00	0.00	0.00	0.00	0.00	0.00	0.00	0.00	0.00	0.00	0.00	0.00
20	(s) Copper (Cu)	g	0.00	0.00	0.00	0.00	0.00	0.00	0.00	0.00	0.00	0.00	0.00	0.00	0.00	0.00	0.00
20	(s) Iron (Fe)	g	0.00	0.00	0.00	0.00	0.00	0.00	0.00	0.00	0.00	0.00	0.00	0.00	0.00	0.00	0.00

Exhibit D-3. (Continued)

BEES Data file B1011A: Generic Concrete Beam, 100% Portland Cement (4KSI)																	
XPORT DIST	FLOW	UNIT	TOTAL	COMP1	COMP2	COMP3	RAW1	RAW2	RAW3	MFG1	MFG2	MFG3	XPORT1	XPORT2	XPORT3	USE1	WASTE1
20	(s) Lead (Pb)	g	0.00	0.00	0.00	0.00	0.00	0.00	0.00	0.00	0.00	0.00	0.00	0.00	0.00	0.00	0.00
20	(s) Manganese (Mn)	g	0.00	0.00	0.00	0.00	0.00	0.00	0.00	0.00	0.00	0.00	0.00	0.00	0.00	0.00	0.00
20	(s) Mercury (Hg)	g	0.00	0.00	0.00	0.00	0.00	0.00	0.00	0.00	0.00	0.00	0.00	0.00	0.00	0.00	0.00
20	(s) Nickel (Ni)	g	0.00	0.00	0.00	0.00	0.00	0.00	0.00	0.00	0.00	0.00	0.00	0.00	0.00	0.00	0.00
20	(s) Nitrogen (N)	g	0.00	0.00	0.00	0.00	0.00	0.00	0.00	0.00	0.00	0.00	0.00	0.00	0.00	0.00	0.00
20	(s) Oils (unspecified)	g	0.00	0.00	0.00	0.00	0.00	0.00	0.00	0.00	0.00	0.00	0.00	0.00	0.00	0.00	0.00
20	(s) Phosphorus (P)	g	0.00	0.00	0.00	0.00	0.00	0.00	0.00	0.00	0.00	0.00	0.00	0.00	0.00	0.00	0.00
20	(s) Sulfur (S)	g	0.00	0.00	0.00	0.00	0.00	0.00	0.00	0.00	0.00	0.00	0.00	0.00	0.00	0.00	0.00
20	(s) Zinc (Zn)	g	0.00	0.00	0.00	0.00	0.00	0.00	0.00	0.00	0.00	0.00	0.00	0.00	0.00	0.00	0.00
20	(w) BOD5 (Biochemical Oxygen	g	15.80	7.04	7.47	1.28	6.25	7.45	1.28	0.11	0.00	0.00	0.68	0.02	0.00	0.00	0.00
20	(w) COD (Chemical Oxygen Dem	g	82.36	59.57	20.40	2.39	52.89	20.26	2.37	0.92	0.00	0.01	5.76	0.14	0.01	0.00	0.00
20	(w) Copper (Cu+, Cu++)	g	0.08	0.00	0.08	0.00	0.00	0.08	0.00	0.00	0.00	0.00	0.00	0.00	0.00	0.00	0.00
20	(w) Suspended Matter (unspec	g	43.64	31.97	9.85	1.81	28.39	9.78	1.80	0.49	0.00	0.01	3.09	0.07	0.01	0.00	0.00
20	Waste (end-of-Life)	kg	1,883.35	0.00	0.00	0.00	0.00	0.00	0.00	0.00	0.00	0.00	0.00	0.00	0.00	0.00	1,883.35
20	E Total Primary Energy	MJ	2,779.14	1,994.61	658.19	126.35	1,904.34	656.30	121.11	12.42	0.00	5.07	77.86	1.88	0.16	0.00	0.00

Exhibit D-4. Bees Life Cycle Inventory Data for Concrete Beam without Blended Cement

BEES Datafile B1011B: Generic Concrete Beam, 100% Portland Cement (4KSI)																	
XPORT DIST	FLOW	UNIT	TOTAL	COMP1	COMP2	COMP3	RAW1	RAW2	RAW3	MFG1	MFG2	MFG3	XPORT1	XPORT2	XPORT3	USE1	WASTE1
20	Water Used (total)	liter	1,690.06	1,043.05	570.94	4.39	999.10	570.02	4.25	6.05	0.00	0.07	37.91	0.92	0.08	71.67	71.67
20	Concrete Beam	Cu yd	1.00	0.00	0.00	0.00	0.00	0.00	0.00	0.00	0.00	0.00	0.00	0.00	0.00	0.00	0.00
20	Installation component 1	kg	65.77	0.00	65.77	0.00	0.00	0.00	0.00	0.00	0.00	0.00	0.00	0.00	0.00	0.00	0.00
20	Main component	kg	1,817.58	1,817.58	0.00	0.00	0.00	0.00	0.00	0.00	0.00	0.00	0.00	0.00	0.00	0.00	0.00
20	Installation component 2	kg	28.57	0.00	0.00	28.57	0.00	0.00	0.00	0.00	0.00	0.00	0.00	0.00	0.00	0.00	0.00
20	Component 4	kg	0.00	0.00	0.00	0.00	0.00	0.00	0.00	0.00	0.00	0.00	0.00	0.00	0.00	0.00	0.00
20	Component 5	kg	0.00	0.00	0.00	0.00	0.00	0.00	0.00	0.00	0.00	0.00	0.00	0.00	0.00	0.00	0.00
20	Component 6	kg	0.00	0.00	0.00	0.00	0.00	0.00	0.00	0.00	0.00	0.00	0.00	0.00	0.00	0.00	0.00
20	(a) Carbon Dioxide (CO2, fos	g	243,685.00	191,547.00	50,991.90	1,146.09	185,379.00	50,863.70	815.22	862.43	0.00	319.85	5,305.62	128.19	11.02	0.00	0.00
20	(a) Carbon Tetrafluoride (CF	g	0.00	0.00	0.00	0.00	0.00	0.00	0.00	0.00	0.00	0.00	0.00	0.00	0.00	0.00	0.00
20	(a) Lead (Pb)	g	0.43	0.01	0.42	0.00	0.01	0.42	0.00	0.00	0.00	0.00	0.00	0.00	0.00	0.00	0.00
20	(a) Mercury (Hg)	g	0.01	0.01	0.00	0.00	0.01	0.00	0.00	0.00	0.00	0.00	0.00	0.00	0.00	0.00	0.00
20	(a) Methane (CH4)	g	278.61	187.66	88.66	2.29	183.56	88.57	1.55	0.57	0.00	0.73	3.52	0.09	0.01	0.00	0.00
20	(a) Nitrogen Oxides (NOx as	g	1,231.00	1,103.86	118.58	8.56	1,027.87	117.07	4.87	13.60	0.00	3.56	62.38	1.51	0.13	0.00	0.00
20	(a) Nitrous Oxide (N2O)	g	6.68	6.28	0.28	0.12	5.53	0.26	0.08	0.03	0.00	0.04	0.73	0.02	0.00	0.00	0.00
20	(a) Particulates (PM 10)	g	0.01	0.01	0.00	0.00	0.01	0.00	0.00	0.00	0.00	0.00	0.00	0.00	0.00	0.00	0.00
20	(a) Sulfur Oxides (SOx as SO	g	555.41	425.95	125.58	3.88	418.19	125.41	2.64	0.71	0.00	1.23	7.06	0.17	0.01	0.00	0.00
20	(s) Aluminum (Al)	g	0.00	0.00	0.00	0.00	0.00	0.00	0.00	0.00	0.00	0.00	0.00	0.00	0.00	0.00	0.00
20	(s) Arsenic (As)	g	0.00	0.00	0.00	0.00	0.00	0.00	0.00	0.00	0.00	0.00	0.00	0.00	0.00	0.00	0.00
20	(s) Cadmium (Cd)	g	0.00	0.00	0.00	0.00	0.00	0.00	0.00	0.00	0.00	0.00	0.00	0.00	0.00	0.00	0.00
20	(s) Carbon (C)	g	0.00	0.00	0.00	0.00	0.00	0.00	0.00	0.00	0.00	0.00	0.00	0.00	0.00	0.00	0.00
20	(s) Calcium (Ca)	g	0.00	0.00	0.00	0.00	0.00	0.00	0.00	0.00	0.00	0.00	0.00	0.00	0.00	0.00	0.00
20	(s) Chromium (Cr III, Cr VI)	g	0.00	0.00	0.00	0.00	0.00	0.00	0.00	0.00	0.00	0.00	0.00	0.00	0.00	0.00	0.00
20	(s) Cobalt (Co)	g	0.00	0.00	0.00	0.00	0.00	0.00	0.00	0.00	0.00	0.00	0.00	0.00	0.00	0.00	0.00
20	(s) Copper (Cu)	g	0.00	0.00	0.00	0.00	0.00	0.00	0.00	0.00	0.00	0.00	0.00	0.00	0.00	0.00	0.00
20	(s) Iron (Fe)	g	0.00	0.00	0.00	0.00	0.00	0.00	0.00	0.00	0.00	0.00	0.00	0.00	0.00	0.00	0.00

Exhibit D-4. (Continued)

BEES Datafile B1011B: Generic Concrete Beam, 100% Portland Cement (4KSI)																	
XPORT DIST	FLOW	UNIT	TOTAL	COMP1	COMP2	COMP3	RAW1	RAW2	RAW3	MFG1	MFG2	MFG3	XPORT1	XPORT2	XPORT3	USE1	WASTE1
20	(s) Lead (Pb)	g	0.00	0.00	0.00	0.00	0.00	0.00	0.00	0.00	0.00	0.00	0.00	0.00	0.00	0.00	0.00
20	(s) Manganese (Mn)	g	0.00	0.00	0.00	0.00	0.00	0.00	0.00	0.00	0.00	0.00	0.00	0.00	0.00	0.00	0.00
20	(s) Mercury (Hg)	g	0.00	0.00	0.00	0.00	0.00	0.00	0.00	0.00	0.00	0.00	0.00	0.00	0.00	0.00	0.00
20	(s) Nickel (Ni)	g	0.00	0.00	0.00	0.00	0.00	0.00	0.00	0.00	0.00	0.00	0.00	0.00	0.00	0.00	0.00
20	(s) Nitrogen (N)	g	0.00	0.00	0.00	0.00	0.00	0.00	0.00	0.00	0.00	0.00	0.00	0.00	0.00	0.00	0.00
20	(s) Oils (unspecified)	g	0.00	0.00	0.00	0.00	0.00	0.00	0.00	0.00	0.00	0.00	0.00	0.00	0.00	0.00	0.00
20	(s) Phosphorus (P)	g	0.00	0.00	0.00	0.00	0.00	0.00	0.00	0.00	0.00	0.00	0.00	0.00	0.00	0.00	0.00
20	(s) Sulfur (S)	g	0.00	0.00	0.00	0.00	0.00	0.00	0.00	0.00	0.00	0.00	0.00	0.00	0.00	0.00	0.00
20	(s) Zinc (Zn)	g	0.00	0.00	0.00	0.00	0.00	0.00	0.00	0.00	0.00	0.00	0.00	0.00	0.00	0.00	0.00
20	(w) BOD5 (Biochemical Oxygen	g	15.69	6.93	7.47	1.28	6.14	7.45	1.28	0.11	0.00	0.00	0.68	0.02	0.00	0.00	0.00
20	(w) COD (Chemical Oxygen Dem	g	81.45	58.66	20.40	2.39	51.98	20.26	2.37	0.92	0.00	0.01	5.76	0.14	0.01	0.00	0.00
20	(w) Copper (Cu+, Cu++)	g	0.08	0.00	0.08	0.00	0.00	0.08	0.00	0.00	0.00	0.00	0.00	0.00	0.00	0.00	0.00
20	(w) Suspended Matter (unspec	g	43.14	31.48	9.85	1.81	27.90	9.78	1.80	0.49	0.00	0.01	3.09	0.07	0.01	0.00	0.00
20	Waste (end-of-Life)	kg	1,883.35	0.00	0.00	0.00	0.00	0.00	0.00	0.00	0.00	0.00	0.00	0.00	0.00	0.00	1,883.35
20	E Total Primary Energy	MJ	2,629.00	1,844.47	658.19	126.35	1,754.19	656.30	121.11	12.42	0.00	5.07	77.86	1.88	0.16	0.00	0.00

Exhibit D-5. Example Calculation Of Impact Metric For Water Usage Related To Fly Ash Substitution In Concrete

	CALCULATION	3 KSI CONCRETE PAVEMENT	NOTE/SOURCES
IMPACTS PER CUBIC YARD CONCRETE			
100% portland cement		266,110 grams per cubic yard of concrete	Values represent impacts related to a 4 KSI concrete beam as characterized in BEES data file B1011A. BEES Version 3.0 Performance Data.
15% coal fly ash		243,685 grams per cubic yard of concrete	Values represent impacts related to building products and pavement as characterized in BEES data file B1 011 B. BEES Version 3.0 Performance Data.
Incremental benefit	[c]=[a]-[b]	22,425 grams per cubic yard of concrete	Represents avoided CO_2, in grams per cubic yard of concrete product substituting 15% coal fly ash for portland cement.
IMPACTS PER U.S. SHORT TON FLY ASH			
lbs cement/yd^3 concrete	[d]	470 lbs cement/cubic yard of concrete	Represents proportion of one cubic yard of concrete made up of cementitious material, given a mix-design or constituent density (Lipiatt, 2002, p. 40).
Percent coal fly ash substitution	[e]	15%	Fifteen percent of cementitious material is replaced with coal fly ash.
lbs/U.S. short ton	[f]	2000 lbs/ton	Conversion factor for pounds to tons.
tons fly ash/yd^3 concrete	[g]=[d]*[e]/[f]	0.0352 tons coal fly ash/cubic yard of concrete	Conversion of quantity of coal fly ash in one cubic yard of concrete from pounds to tons.
unit impact	[h]=[c]/[g]	636,170 grams per ton of coal fly ash substituted for cement	Represent unit impact values for CO_2 (in grams), based on substitution of one ton of coal fly ash for 1 ton portland cement in concrete.

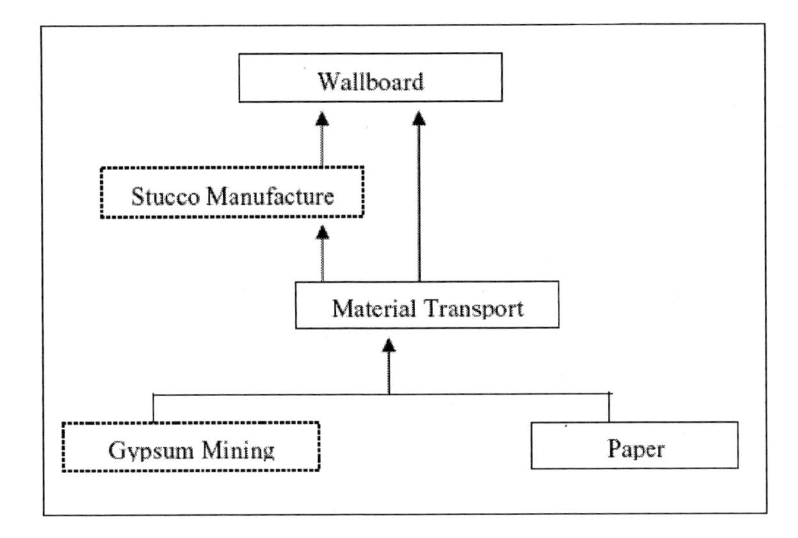

Exhibit D-6. Virgin Gypsum Wallboard Manufacture

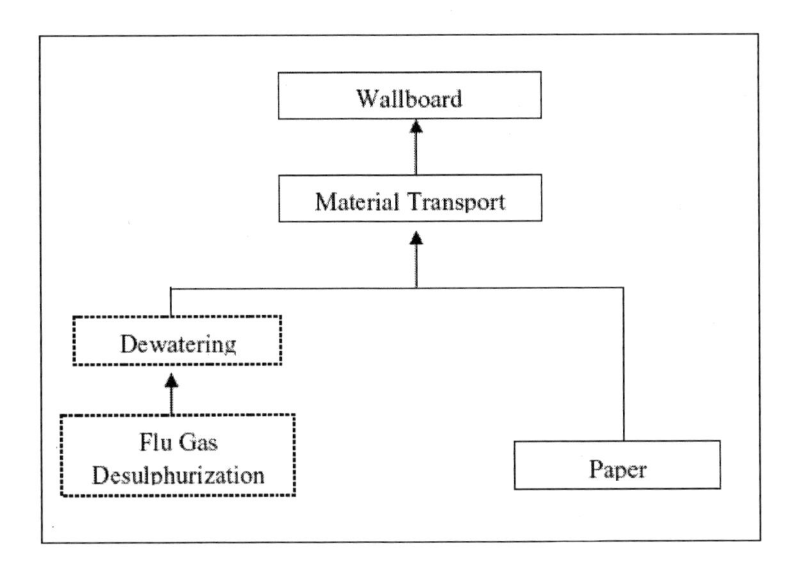

Exhibit D-7. FGD Gypsum Wallboard Manufacture

Calculation of Unit Impacts

To illustrate the methodology used to calculate the unit impact values from the BEES life cycle inventory data, we present a sample calculation of CO_2 reductions resulting from the substitution of one ton of fly ash for finished portland cement in concrete (see Exhibit D-6).

The process outlined in Exhibit D-6 is repeated for each of the environmental metrics evaluated in this analysis using the environmental performance data reported in BEES. For each environmental metric, this yields an estimate of the benefit of one ton of fly ash replacing finished portland cement in concrete.

SIMAPRO ANALYSIS OF FGD GYPSUM IN WALLBOARD

We calculate the unit impacts of using FGD gypsum in place of virgin gypsum stucco in wallboard as the difference in impacts between wallboard made with 100% virgin gypsum and wallboard made with 100% FGD gypsum. Exhibits D-6 and D-7 show the lifecycle stages in the production of wallboard with 100% virgin gypsum and 100% FGD gypsum, respectively. The boxes with dashed lines represent life cycle stages that are unique to virgin or FGD gypsum.

As shown in Exhibits D-6 and D-7, by replacing virgin gypsum stucco with FGD gypsum, gypsum mining and stucco manufacture can be avoided but a dewatering step is added to the lifecycle. We model these impacts as one ton of avoided "stucco" manufacture in SimaPro.[75] Stucco is the term used in SimaPro to describe the gypsum material used in wallboard. We selected the EcoInvent data set because it includes gypsum mining but also includes the processing of gypsum for use in wallboard (i.e., burning of gypsum and milling of stucco for use in gypsum wallboard). Thus, this dataset includes all the processes that would be avoided if an equivalent quantity of FGD gypsum were used in place of stucco in wallboard. The production of FGD gypsum through the coal combustion process is not modeled, as discussed in Chapter 4. In addition, this analysis assumes that FGD gypsum dewatering occurs via holding ponds and that the environmental impacts are negligible.[76] This analysis also does not model transport distance; we assume FGD gypsum would have the same transport distances to the construction site as virgin gypsum.[77] Thus avoided gypsum mining and avoided processing of gypsum into stucco, as represented by the EcoInvent stucco manufacturing data set, are the only lifecycle stages modeled in SimaPro.

Exhibit D-8 presents the assumed life cycle system boundaries for the EcoInvent stucco manufacturing data set. The cut-off node for the process flow tree depicted in Exhibit D-8 is set to 0.5% so that the entire tree could be viewed. Thus, this process tree lists the flows associated with 99.5% of the total life cycle impacts for one ton of stucco manufacture.[78] The numbers that appear in the bottom left-hand corner of each box in the tree are partition factors used by SimaPro and are not central to this analysis.

The EcoInvent dataset for stucco manufacture is the only stucco dataset available in SimaPro, but because it reflects Swiss manufacturing processes and electricity mix, we modified the data for stucco manufacture to reflect the average U.S. electricity mix. We made this modification by substituting the electricity used to make stucco from gypsum, as well as the electricity used further down the production chain in gypsum mining, with the Franklin data set for average U.S. electricity mix titled "Electricity avg. kWh USA". The Franklin data set includes the fuel consumption associated with the generation and delivery of an average kilowatt-hour in the USA using average USA technology in the late 1990's. While we did not substitute the U.S. electricity mix at points further down the stucco production chain, the stucco manufacturing and gypsum mining processes account for the majority of electricity use in this analysis.[79]

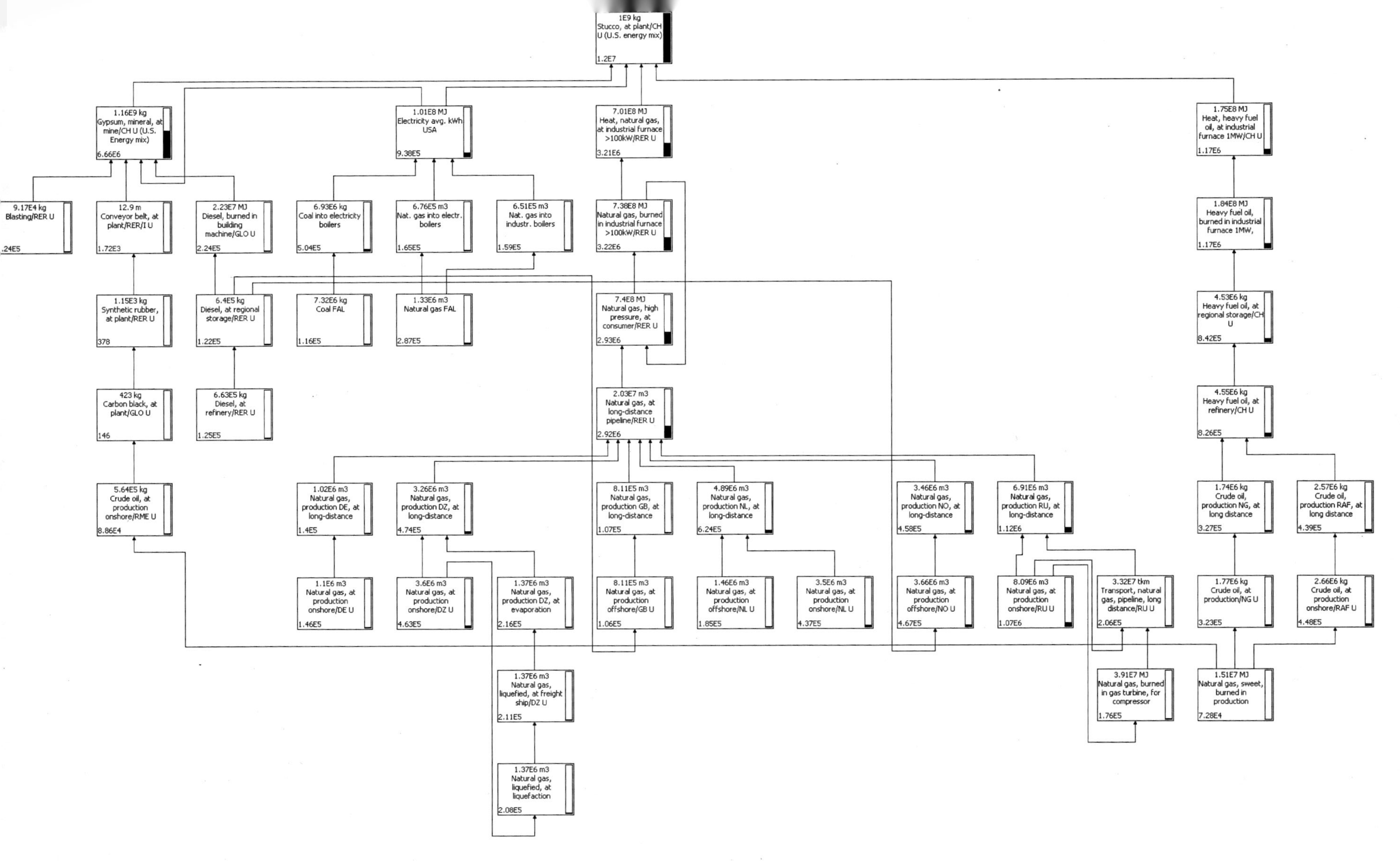

Exhibit D-8. Life Cycle System Boundaries For Stucco Manufacture, 0.5% Cut-Off

Exhibit D-9. Simapro LCI Data for One Ton Stucco, at Plant

SUBSTANCE	COMPARTMENT	UNIT	STUCCO, AT PLANT/CH U (U.S. ENERGY MIX)	CONVERTED TO BEES UNITS		NOTES
Gypsum, in ground	Raw	tn.lg	1.046899978			
Total Energy				MJ	12,582.66	
Non-Renewable Energy				MJ	12,568.97	
Coal, 26.4 MJ per kg, in ground	Raw	kg	6.756131931	MJ	178.36	
Coal, brown, in ground	Raw	g	315.4661176	MJ	7.61	*converted using the EIA's Energy Calculator[a]
Coal, hard, unspecified, in ground	Raw	g	412.5766501	MJ	9.96	*converted using the EIA's Energy Calculator[a]
Gas, natural, 46.8 MJ per kg, in ground	Raw	kg	1.156160311	MJ	54.11	
Gas, natural, in ground	Raw	m3	20.95771832	MJ	799.32	*converted using the EIA's Energy Calculator[a]
Oil, crude, 42 MJ per kg, in ground	Raw	g	268.9543393	MJ	11,296.08	
Oil, crude, in ground	Raw	kg	4.981418424	MJ	223.53	*converted using the EIA's Energy Calculator[a]
Renewable Energy				MJ	13.69	
Energy, from hydro power	Raw	MJ	9.697825819	MJ	9.70	
Energy, gross calorific value, in biomass	Raw	kJ	454.7219368	MJ	0.45	
Energy, kinetic, flow, in wind	Raw	kJ	235.1774187	MJ	0.24	
Energy, potential, stock, in barrage water	Raw	MJ	3.294986059	MJ	3.29	
Energy, solar	Raw	kJ	3.740125096	MJ	0.00	
Fresh Water Use				**liter**	**14,214.60**	
Water, cooling, unspecified natural origin/m3	Raw	dm3	58.63469673	liter	58.63	
Water, lake	Raw	cm3	149.6500813	liter	0.15	
Water, river	Raw	cu.in	704.9737212	liter	11.55	
Water, turbine use, unspecified natural origin	Raw	m3	14.11964941	liter	14,119.65	

Exhibit D-9. (Continued)

SUBSTANCE	COMPARTMENT	UNIT	STUCCO, AT PLANT/CHU (U.S. ENERGY MIX)	CONVERTED TO BEES UNITS		NOTES
Water, unspecified natural origin/m3	Raw	dm3	22.25850268	liter	22.26	
Water, well, in ground	Raw	cu.in	143.8279244	liter	2.36	
Carbon dioxide, biogenic	Air	g	39.8286174			
Carbon dioxide, fossil	Air	kg	77.75423811	g	77,754.24	
Carbon Monoxide				g	39.059865	
Carbon monoxide	Air	g	9.12143624			
Carbon monoxide, fossil	Air	g	29.9384285			
Lead	Air	mg	28.76417316	g	0.03	
Mercury	Air	μg	976.1346395	g	0.00	
Methane				g	175.51	
Methane	Air	g	39.14177696			
Methane, fossil	Air	g	136.3669305			
Nitrogen oxides	Air	g	168.024936	g	168.02	
Ozone	Air	mg	11.45337237	g	0.0114534	
Particulates < PM10				**g**	**520.9278**	
Particulates, < 10 um	Air	g	3.45821205			
Particulates, < 2.5 um	Air	g	90.43917261			
Particulates, > 2.5 um, and < 10um	Air	g	427.0304396			
Particulates, > 10 um	**Air**	**kg**	**1.194254106**	**g**	**1,194.25**	
Particulates, unspecified	**Air**	**g**	**17.10845282**	**g**	**17.11**	
Sulfur oxides	**Air**	**g**	**139.1401881**	**g**	**139.1402**	
BOD5, Biological Oxygen Demand	**Water**	**g**	**21.86848366**	**g**	**21.87**	
COD, Chemical Oxygen Demand	**Water**	**g**	**24.71218062**	**g**	**24.71**	
Copper, ion	**Water**	**mg**	**23.66432149**	**g**	**0.02**	
Lead	**Water**	**mg**	**7.909309986**	**g**	**0.01**	
Mercury	**Water**	**μg**	**306.2803191**	**g**	**0.00**	
Suspended solids, unspecified	**Water**	**g**	**23.59769783**	**g**	**23.60**	
Selenium	**Water**	**μg**	**286.8551494**	**g**	**0.00**	
Waste, solid	**Waste**	**kg**	**3.115903858**	**kg**	**3.12**	

Notes: a. Accessed at: http://www.eia.doe.gov/kids/energyfacts/science/energycalculator.html#coalcalc.

SimaPro Life Cycle Inventory Data

Exhibit D-9 presents the SimaPro lifecycle inventory data for stucco manufacture for the same metrics evaluated in the BEES analysis.

In order to easily compare the results of the FGD gypsum analysis with those of the fly ash analysis, it was necessary to convert the environmental metrics reported by SimaPro into the same units that are reported by BEES. For most metrics, this required only a simple conversion between different units of mass. In the case of energy use, however, SimaPro reports quantities of various fossil fuels consumed whereas BEES reports energy consumed in megajoules. To convert the fossil fuel quantities reported by SimaPro into equivalent energy content in megajoules, we relied on the Energy Information Administration's Coal, Natural Gas, and Crude Oil Conversion Calculators.[80]

In addition, SimaPro does not report a single "water use" metric, as is done in BEES, but breaks out fresh water use by origin (e.g., lake, river, well, etc.) and application (cooling, turbine, etc). We sum the following metrics (converted to liters) to obtain the water use figure in the FGD gypsum analysis: 1) Water, cooling, unspecified natural origin/m^3, 2) Water, lake, 3) Water, river, 4) Water, turbine use, unspecified natural origin, 5) Water, unspecified natural origin/m^3, and 6) Water, well, in ground.

APPENDIX E. POTENTIAL IMPACTS OF ALLOCATION OF LCI RESULTS TO CCPS

While the background literature (ISO framework, etc.) are relatively consistent in their discussion that only co-products should share allocation of input and output system flows, this rule leaves out the consideration of current and future "waste streams" that have beneficial use potential, or market value that suggests that they may be usefully treated as co-products.

This observation is inspired by the need to consider the net impacts of CCPs in electricity generation when looking at the life cycle impacts associated with beneficial use. While there are beneficial substitutions possible of fly ash for cement, FGD gypsum for virgin gypsum, etc., it is possible that if the CCPs were in fact treated as co-products instead of as wastes, that there would be non-negligible inputs and outputs from coal-fired electricity generation that merited attention when estimating net impacts. In this section we consider some hypothetical macro-level scenarios for coal-fired power generation, as well as macro-level flows of several key CCPs. These scenarios are then applied in an assessment of implications of allocating the environmental effects of power generation to both the energy product and the CCPs.

Traditional LCA allocation rules suggest that product and co-product allocation by economic value, mass, energy, etc., are all legitimate methods – there is no single approach to allocating that is correct. For the first illustration, we show an approximate economic value based allocation and the resulting effects for CCPs, followed by a prospective mass-based method.

Table E-1. Optimistic Scenario of CCP Market Values (ACAA 2005, USGS 2006)

CCP	MARKET VALUE (PER TON)	CCP PRODUCTION (MILLION TONS)	TOTAL MARKET VALUE (ABSOLUTE UPPER BOUND)
Fly Ash	$45	71	$3.00 billion
FGD Gypsum	$31	12	$0.37 billion
Bottom Ash	$8	18	$0.14 billion
Total --	-------	101	$3.51 billion

Economic Allocation

The electricity industry has about $300 billion per year in gross revenues. Roughly 50% of generation is coal-fired at the national level. Even though the costs and revenues per kilowatt hour vary across generation types, and the total value includes generation, transmission, and distribution, for simplicity we assume that there is 50 percent, or $150 billion of revenues from coal-fired power generation. If we were to adjust for the variations in price per kilowatt hour, this value would likely be closer to $100 billion from coal-fired generation, as coal represents a lower-cost form of energy production.

From ACAA (2005) and USGS (2006), we consider the upper bound economic value of various CCPs, using both the high end of estimated market value for the CCPs, as well as the high end estimate of CCPs produced, and not the quantity used. Table E-1 summarizes these results for the top three CCPs in terms of market value and production.

Summing the total value of these three products yields $3.5 billion. Even this optimistic, upper-bound estimate is only 2% to 4% of the value of the electricity produced , if considered as shares of the total economic value of the product (electricity) and co-products (CCPs) created by coal-fired power plants. Of course, the values in Table E-1 are highly optimistic, and USGS estimates that fly ash market value ranges from $0-45/ton, and bottom ash from $4-8 per ton. Thus, the actual economic value allocation would likely be significantly smaller, probably less than one percent. As these were the "best case" allocation results, it seems that allocating by economic value would lead to negligible results.

MASS-BASED ALLOCATION

The example above is straightforward in demonstrating that economic allocation is possible and feasible, but leads to negligible results. Another alternative in LCA is to use mass-based allocation of impacts from products and co-products of a process. In the case of CCPs from coal-fired power generation, the product is electricity, which has no mass, which means it is impossible to purely allocate by mass. However, as an illustration, we consider the allocation results assuming that the electricity generated is completely tied to the combustion of coal, which has known mass. This is a simplifying but fair assumption since there are few other significantly large mass based inputs into coal combustion processes (process water is generally reused and returned).

If the mass-based allocation were considered as such, and thinking again at the macro-level of all coal- fired power plants, there are about one billion tons of coal used as input. As summarized in Table E-1, there are about 100 million tons of the top three market value CCPs, and about 120 million tons total CCPs, generated per year by the power plants. Thus, CCPs represent about 12% of the mass, with individual mass allocations of about 7% for fly ash, one percent for gypsum, and 0.2% for boiler slag. Thus, in this hypothetical example, the mass-based allocations would, in fact be much larger than the economic value allocations, but still generally a small percentage of total "mass" production. Further, considering the other major mass flows in the plant, these numbers may, in fact, be smaller. Another caveat is that not all CCPs produced are beneficially used. Thus, the mass allocations may converge back to the shares estimated above for economic allocation.

SAMPLE CALCULATIONS

Given the substitution of CCPs for virgin materials production, and the potential effects of allocating some of the environmental flows of electric power generation to CCPs, we investigated what the comparative net effects would be if the estimated low range of mass or dollar based allocations for CCPs (of coal-fired generation) were compared to the avoided emissions from coal fly ash and FGD gypsum beneficial use. For simplicity we consider CO_2 and SO_2 emissions only.

In 2005 there were 2.5 billion metric tons of CO_2 emitted in all electricity generation.[81] The latest data available from DOE that show emissions by generation type (1999) suggests that 80% of CO_2 emissions come from coal-fired generation, with an effective emissions factor of 2 lbs/kWh (or roughly 1 short ton/MWh).[82] Assuming the same emissions rate, the 1.5 billion MWh of coal-fired generation in 2005 would have emitted 1.5 billion tons of CO_2.[83] Given the published 2005 emissions of 2.5 billion metric tons CO_2, this 1.5 billion metric ton estimate is less than 60% of CO_2 emissions, and thus may be low.

From our overview of potential allocation values for CCPs from coal-fired generation, we estimated that the percent allocations would be, in sum, on the order of about one percent. If we allocated one percent of CO_2 emissions from coal-fired electricity generation to the CCPs, then about 15 million (short) tons of CO_2 would be allocated to their "production." In comparing this one percent allocated value to the CO_2 benefits estimated separately by BEES and Simapro, we see that the avoided portland cement and virgin gypsum use accounts for about 11.5 million short tons of CO_2 emission benefits. Similarly, for SO_2, coal- fired generation leads to most electricity generation emissions, which total about 10 million metric tons per year.[84] One percent of this number is about 100,000 metric tons SO_2, though this includes emissions from all Conventional Power Plants and Combined-Heat-and-Power Plants and therefore overstates the impact of coal combustion. However, Exhibit 5-3 in Chapter 5 of this book estimates avoided cement and gypsum manufacturing SO2 emissions to be 26,000 short tons (23.9 million metric tons or 23.9 billion grams), suggesting that SO_2 emissions reductions associated with beneficial reuse are small when compared with the allocated emissions impacts associated with energy production from coal.

As indicated in our high-end CO_2 and SO_2 examples presented above, allocated emissions from primary production (i.e., coal combustion) may occasionally be greater than the documented benefits of beneficial use for some metrics. However, it is important to note that this allocation procedure reflects an accounting procedure designed only to more accurately apportion emission impacts across co-products. It can be correctly interpreted as an indication that the beneficial use of CCPs may not be an efficient method for reducing overall emissions of CO_2 and SO_2 to the environment. However, the actual CO_2 and SO_2 emissions avoided from the beneficial use of coal fly ash and FGD gypsum remain positive, as reported.

While this analysis has focused only on CO_2 and SO_2 (there are similar emissions from coal-fired generation of NOx, PM_{10}, etc.), it demonstrates the type of framework that could be in place to help assess the efficiency of beneficial use. It is likely that within such a framework that life cycle inventory data would be greater for one effect and lower for others, rather than a vector dominance situation. Thus, appropriate weighting methods should be identified to help balance the overall perceived benefit of such substitutions. EPA's TRACI model and the BEES model itself could serve to normalize and weight preferences of environmental flows against each other to lead to singular assessments of results.

SUMMARY

LCA allows for the allocation of input and output flows across the life cycle to the various products and co-products of processes and systems. However CCPs are generally considered waste, and not co- products of power generation. Even if they were considered co-products, the allocated input and output flows from coal-fired generation would associate only very small flows to the CCPs relative to the electricity produced. For this reason, and because our assessment here represents a high-end screening analysis, we do not include either an economic or mass-based allocation of coal combustion impacts to fly ash or FGD gypsum in our presentation of extrapolated findings in chapter 5.

REFERENCES

ACAA (American Coal Ash Association). (2006). 2006a. Dave Goss. Personal Communication. March.

ACAA (American Coal Ash Association). (2006). 2006b. Dave Goss. Personal Communication. April.

ACAA (American Coal Ash Association). (2006). 2006c. Dave Goss. Personal Communication. May.

ACAA (American Coal Ash Association). (2006). 2006c. Dave Goss. Personal Communication. November.

ACAA (American Coal Ash Association). undated. Accessed at: www.acaa-usa.org.

American Coal Ash Association. "2005 Coal Combustion Product (CCP) Production and Use Survey," accessed at: http://www.acaa-usa.org/PDF/20045_CCP_
Production_and_Use_Figures_Released_by_ACAA.pdf.

American Coal Ash Association, "Frequently Asked Questions," accessed at: www.acaa-usa.org

American Coal Foundation. (2003a). "*All About Coal: Fast Facts About Coal.*" Accessed at: http://www.teachcoal.org/aboutcoal/articles/fastfacts.html.

American Coal Foundation. (2003b). "*Coal's Past Present and Future*." Accessed at: http://www.teachcoal.org/aboutcoal/articles/coalppf.html.

American Foundry Society (AFS). (2007). "*Foundry Industry Benchmarking Survey*," August.

Apul, Define. (2005). University of Toledo. "*2nd Quarterly Report for EPA-OS WER Innovation Pilot FY05: Development of a Beneficial Reuse Tool for Managing Industrial Byproduct*,".

Caltrans (California Department of Transportation). (2006). Tom Pyle. Personal Communication. November 2006.

Center for Energy and Economic Development. undated. "Growing Demand." Accessed at: http://www.ceednet.org/ceed/index.cfm?cid=7500,7582.

DOE (U.S. Department of Energy). (2007). Energy Information Administration, "Energy Basics 101," http://www.eia.doe.gov/basics/energybasics101.html, August.

DOE (U.S. Department of Energy). (2007). Energy Information Administration, "U.S. Coal Consumption by End- Use Sector," http://www.eia.doe.gov/cneaf/coal June.

DOT (U.S. Department of Transportation). (2006). John D'Angelo. Personal Communication. March 2006.

Edgar G. Hertwich, Sarah F. Mateles, William S. Pease & Thomas E. McKone. (2001). *Environmental Toxicology and Chemistry, 20*(4), 928-939.

Electric Power Research Institute. (1999). Environmental Focus: Flue Gas Desulfurization By-Products. BR-1 14239.

Electric Power Research Institute. (2006). "Coal Combustion Product Use." Accessed at: http://www.epri.com/Portfolio/product.aspx?id=2065&area=50.

Energy and Environmental Research Center. (1999). "Barriers to the Increased Utilization of Coal Combustion/Desulfurization By-Products by Government and Commercial Sectors--Update 1998," EERC Topical Report DE-FC21-93MC-30097--79, July.

Energy & Environmental Research Center. (2006). University of North Dakota, "*Review of Florida Regulations, Standards, and Practices Related to the Use of Coal Combustion Products: Final Report*." University of North Dakota. April 2006. Accessed at: http://www.undeerc.org/carrc/Assets/TB-FLStateReviewFinal.pdf.

Energy & Environmental Research Center. (2005). University of North Dakota, "*Review of Texas Regulations, Standards, and Practices Related to the Use of Coal Combustion Products: Final Report*," January, Accessed at: http://www.undeerc.org/carrc /Assets/TXStateReviewFinalReport.pdf.

EPA. (U.S. Environmental Protection Agency). (1999). Report to Congress: Wastes from the Combustion of Fossil Fuels. Vol. II. EPA-530-R-99-010, March 1999.

EPA. (U.S. Environmental Protection Agency). (2003). "*Underground Storage Tanks (UST) Cleanup & Resource Conservation & Recovery Act (RCRA) Subtitle C Program Benefits, Costs, & Impacts (BCI) Assessments: An SAB Advisory*." EPA-SAB-EC-ADV-03-001

EPA. (U.S. Environmental Protection Agency). (2005). Using Coal Ash in Highway Construction: A Guide to Benefits and Impacts. EPA-530-K-05-002, April 2005.

EPA. (U.S. Environmental Protection Agency). (2006). "About C^2P^2." Accessed at: http://www.epa.gov/epaoswer/osw/conserve/c2p2/about/about.htm.

EPA. (U.S. Environmental Protection Agency). (2006). "Boiler Slag." Accessed at: http://www.epa.gov/epaoswer/osw/conserve/c2p2/about/about.htm.

EPA. (U.S. Environmental Protection Agency). (2006)."Advisory on EPA's Superfund Benefits Analysis." EPA-SAB-ADV-06-002.

Glenn, J., Goss, D. & Sager, J. Undated. "C^2P^2 --Partnership Innovation."

Hendrickson, Chris, Lester Lave & Scott Matthews, H. (2006). *Environmental Life Cycle Assessment of Goods and Services*, Resources for the Future: Washington, DC, 2006.

Hassett, David J., Debra F. Pflughoeft-Hassett, Dennis L. Laudal & John H. Pavlish. (1999). Mercury Release from Coal Combustion ByProducts to the Environment.

Lumia, Deborah, Kristin Linsey & Nancy Barber. (2000). United States Department of the Interior, United States Geological Survey, Summary of Water use in the United States.

Miller, Cheri . Gypsum Parameters. Presentation at WOCA Short Course: "*Strategies for Development of FGD Gypsum Resources*."

Portland Cement Association. (2006). "FAQ: Cement Supply Shortage." Accessed at: http://www.cement.org/pca/shortageQA.asp.

Price, Jason & Mark Ewen. (2006). Industrial Economics, Inc. Memorandum to Lyn Luben, EPA, "*Impact of CAIR and CAMR on the Quantity and Quality of Coal Combustion Fly Ash Generated by Affected Facilities*." December 1.

Stein, Antoinette. (2006). State of California Department of Health Services. Personal Communication, June 2006.

Schwartz, Karen D. (2003). "The Outlook for CCPs," *Electric Perspectives*, July/August 2003.

U.S. Department of Energy, National Energy Technology Laboratory, "General Summary of State Regulations," accessed at: http://www.netl.doe.gov/E&WR/cub/states/select_state.html.

US EPA. (2005). "Characterization of Building-Related Construction and Demolition Debris in the United States" and "Characterization of Road-related Construction and Demolition Debris in the United States,". (Note that these documents were preliminary at the time of this chapter and were undergoing review).

US EPA. (2005). "Draft National Priority Trends Report (1999-2003) Fall," as reported in the NPEP GPRA 2008 database of TRI data from 1998-2003.

US EPA. (2003). "Municipal Solid Waste in the United States: Data Tables," Table 1, accessed on October 26, 2006 at:___http://www.epa.gov/epaoswer/non-hw/muncpl/pubs/03data.pdf

USGS, "Mineral Commodities Summary. (2004). Construction Sand and Gravel," accessed at: http://minerals.usgs.gov/minerals/pubs/commodity USGS, "Mineral Commodities Summary 2005: Lime," accessed at: http://minerals.usgs.gov/minerals/pubs/commodity/lime/lime_myb05.pdf

USGS, "Mineral Commodities Summary (2006). Gypsum," accessed at: http://minerals.usgs.gov/minerals/pubs/commodity

USGS, "Mineral Commodities Summary. (2006). Cement," accessed at: http://minerals .usgs.gov/minerals/pubs/commodity

We Energies. (2007). Tom Jason. *Personal Communication*. November 2006.

Wisconsin Department of Natural Resources. (2006). Bizhan Sheikholeslami. November 2000.

Western Region Ash Group, "Applications and Competing Materials, Coal Combustion Byproducts," accessed at: http://www.wrashg.org/compmat.htm.

Boyd, James, (2004). "What's Nature Worth? Using Indicators to Open the Black Box of Ecological Valuation," *Resources.*

Boyd, James & Lisa Wainger, (2002). "Landscape Indicators of Ecosystem Service Benefits," *American Journal of Agricultural Economics, 84.*

Graedel, T.E., & B.R. Allenby, (1995). *Industrial Ecology*, Prentice Hall: Englewood Cliffs NJ.

Hendrickson, Chris, Lester Lave, & Scott Matthews, H. (2006). *Environmental Life Cycle Assessment of Goods and Services*, Resources for the Future: Washington, DC.

Krupnick, Alan & Dallas Burtraw, (1997). "Social Costs of Electricity: Do the Numbers Add Up?" *Resources and Energy, 18*, 4, 423-466.

Porter, Richard. (2002). *The Economics of Waste*, Resources for the Future: Washington, DC,.

Portney, Paul. (1993). "The Price Is Right: Making Use of Life Cycle Analyses," *Issues in Science and Technology, 10*, 2 Winter, 69-75.

Smith, V. Kerry & Ray Kopp. (1996). *Valuing Natural Assets: The Economics of Natural Resource Damage Assessment*, (Resources for the Future: Washington, DC.

End Notes

[1] U.S. EPA, "About C2P2," accessed at http://www.epa.gov/epaoswer/osw/conserve/c2p2/about/about.htm.

[2] The ACAA survey is administered to both ACAA members and non-members. ACAA members account for approximately 40 percent of private power generation. Not all survey recipients complete the survey each year. ACAA extrapolates survey respondent data to the entire coal-fired electricity generation industry.

[3] It is unclear from the documentation provided for BEES what impacts (e.g. virgin materials extraction, plant infrastructure, etc.) are modeled for portland cement production. For this reason, it is not possible to explain the differences in unit impact results between the FGD gypsum and fly ash analysis.

[4] Based on the assumption that an average residential customer uses 938 kilowatt-hours per month. Department of Energy, Energy Information Administration, "Energy Basics 101,"_http://www.eia.doe.gov/basics/energybasics101.html, accessed August 30, 2007.

[5] Based on 2000 USGS per capita water use estimate of 1,430 gallons per day. Lumia et al., United States Department of the Interior, United States Geological Survey, Summary of Water Use in the United States, 2000.

[6] EIO-LCA models impacts at the sector level using NAICS codes but an individual NAICS code does not exist for the wallboard manufacturing sector.

[7] The RCC is an EPA initiative that seeks to identify and encourage innovative, flexible, and protective ways to conserve natural resources and energy. Specifically, the RCC is a cross-Office program that assists in developing voluntary programs that promote the source reduction, reuse, and recycling of materials.

[8] U.S. EPA, "About C2P2," accessed at http://www.epa.gov/epaoswer/osw/conserve/c2p2/about/about.htm.

[9] As of the writing of this chapter, we were unable to locate data estimating the quantities of CCPs attributable only to the electric power industry; however, since the coal power industry consumes approximately 92 percent of all U.S. coal, it is reasonable to assume that significant majority of CCPs result from the burning of coal at coal-fired power plants. Department of Energy, Energy Information Administration, "U.S. Coal Consumption by End-Use Sector,"_http://www.eia.doe.gov/cneaf/coal/quarterly/html/t28p01p1.html, June 2007.

[10] FGD gypsum has the same chemical structure as naturally occurring gypsum (calcium sulfate dehydrate).

[11] Electric Power Research Institute. 1999. Environmental Focus: Flue Gas Desulfurization By-Products. BR-114239

[12] Cenospheres range in size from 20 to 5000 microns.

[13] The ACAA survey is administered to both ACAA members and non-members. ACAA members account for approximately 40 percent of private power generation. Not all survey recipients complete the survey each year. ACAA extrapolates survey respondent data to the entire coal-fired electricity generation industry. To the extent that other coal-burning industries are not represented in the ACAA sample, the survey may underestimate the quantity of CCPs generated and/or beneficially used.

[14] The quantity of cenospheres generated is not reported by ACAA so the 58 percent estimate could be higher if cenospheres were included.

[15] Personal communications with Dave Goss, ACAA and Tom Janson, WE Energies, November 27, 2006.

[16] The quantity of stockpiled fly ash that is available for beneficial use is unclear. The chemical composition of fly ash varies depending on the type of coal used, and only two types of fly ash--class C fly ash and class F fly ash—meet the ASTM technical requirements for concrete. It is unclear how much of the estimated 100-500 million tons of stockpiled fly ash falls into one of these classes. In addition, exposure to moisture or contamination in the stockpiles can limit the beneficial use options of Class C ash, though, this is not a concern with Class F ash. Information on these standards can be found at http://www.astm.org.

[17] EPA. 1999. "Report to Congress: Wastes from the Combustion of Fossil Fuels." Vol. II. EPA-530-R-99-010, March 1999.

[18] Relatively minor applications comprise the remaining 20 percent of CCP beneficial uses. These applications include use such as soil stabilizers, mineral filler in asphalt, and mine reclamation.

[19] Fly ash is technically a pozzolanic, not a cementitious material. A cementitious material, such as portland cement, is one that hardens when mixed with water. A pozzolanic material will also harden with water but only after activation with an alkaline substance such as lime. The combination of portland cement and water in concrete mixtures creates two products: a durable binder that "glues" concrete aggregates together and free lime. Fly ash reacts with this free lime to create more of the desirable binder.

[20] Personal communication with Tom Pyle, Caltrans, November 2006.

[21] EPA. "Boiler Slag," accessed at: http://www.epa.gov/epaoswer/osw/conserve/c2p2/about/about.htm.

[22] American Coal Ash Association. "2005 Coal Combustion Product (CCP) Production and Use Survey," accessed at: http://www.acaausa.org/PDF/20045_CCP_Production_and_Use_Figures_Released_by_ACAA.pdf. and "2004 Coal Combustion Product (CCP) Production and Use Survey," accessed at: http://www.acaa-usa.org/PDF/2004_CCP_Survey(9-9-05).pdf.

[23] More efficient furnace types that use pulverized coal are replacing the cyclone and slag-tap furnaces that typically produce boiler slag. The replacement of these boiler types is decreasing the available supply of boiler slag. EPA. "Boiler Slag," accessed at: http://www.epa.gov/epaoswer/osw/conserve/c2p2/about/about.htm.

[24] American Coal Foundation, "All About Coal: Fast Facts About Coal," accessed at: http://www.teachcoal.org/aboutcoal/articles/fastfacts.html.

[25] Center for Energy and Economic Development, "Growing Demand," accessed at: http://www.ceednet.org/ceed/index.cfm?cid=7500,7582.

[26] American Coal Foundation, "Coal's Past Present and Future," accessed at: http://www/teachcoal.org/aboutcoal/asrticles/coalppf.html.

[27] Personal communication with Dave Goss, American Coal Ash Association, April 2006.

[28] Personal communication with Dave Goss, American Coal Ash Association, April 2006.

[29] Personal communication with Dave Goss, American Coal Ash Association, May 2006.

[30] Note that the transfer of CCPs from generators to users may lead to potential cost savings for the generators. It may be possible for generators to shift liability (and related costs) associated with CCPs to users of the product. The law on this matter is not well-defined and needs to be clarified to determine the magnitude of any potential cost savings.

[31] Note that the "price" of CCPs represents how much an end-user would pay for the product.

[32] Energy & Environmental Research Center, University of North Dakota, "Review of Florida Regulations, Standards, and Practices Related to theUse of Coal Combustion Products: Final Report," April 2006, accessed at: http://www.undeerc.org/carrc/Assets/TB-FLStateReviewFinal.pdf.

[33] Some states, such as Wisconsin, have set up regulatory schemes designed to speed up the approval process for products using beneficial use materials such as CCPs. Currently, Wisconsin requires initial leachate testing of the material to be beneficially used, which leads to a specific rating. Materials that fall into a standard rating class are automatically approved for specific uses. For example, material found to meet drinking water standards can be used in any application, whereas material found to have a moderate level of contamination, may only be approved for encapsulated uses. Users are also required to submit annual reports demonstrating testing of CCPs. Personal communication with Bizhan Sheikholeslami, Wisconsin Department of Natural Resources, November 2006.

[34] Hassett, David J., Debra F. Pflughoeft-Hassett, Dennis L. Laudal, and John H. Pavlish. 1999. Mercury Release from Coal Combustion ByProducts to the Environment.

[35] Personal Communication with Antoinette Stein, State of California Department of Health Services, June 2006.

[36] Personal communication with Tom Pyle, Caltrans, November 2006.

[37] Strong demand for cement in the U.S. is a result of both increased domestic construction activity and strong demand by growing foreign economies (especially China).

[38] Portland Cement Association. "FAQ: Cement Supply Shortage," accessed at: http://www.cement.org/pca/shortageQA.asp. We contacted two industry experts, Dave Goss of the American Coal Ash Association and Barry Deschenaux of Holcim Cement, to elaborate on this trend (increased use of coal ash in cement due to domestic and foreign demand), but neither was able to provide more detailed information on the extent to which this might occur in the future.

[39] Mercury-CCP dialogue meeting summary document, Final Draft, 1/14/08

[40] Energy and Environment Research Center (EERC), April 2006. Review of Florida Regulations, Standards, and Practices Related to the Use of Coal Combustion Products. 2006-EERC-04-03. University of North Dakota, Grand Forks, ND.

[41] Energy and Environment Research Center (EERC), January 2005. Review of Texas Regulations, Standards, and Practices Related to the Use of Coal ombustion Products. 2005-EERC-01-01. University of North Dakota, Grand Forks, ND.

[42] Electric Power Research Institute, "Environmental Focus: Flue Gas Desulfurization By-Products," 1999.

[43] Personal communication with David Goss, American Coal Ash Association, March 2006.

[44] For example, a beneficial use that has a significant impact on raw material demand (e.g., for virgin aggregate) and on electricity demand may ultimately affect the local prices of both energy and raw materials as demand for these commodities drops. The price changes could, in turn, result in other changes in practice (e.g., decisions on the part of other purchasers to buy more aggregate or changes in use patterns of electricity). These impacts would likely have some impact on the net change in environmental impacts measured in the LCA.

[45] Life cycle analysis can incorporate impact assessments using a range of different methods that can, at a minimum, provide comparative descriptions of the types of damage likely associated (or avoided by) the system change. However, valuation of specific impacts (e.g., health impacts from air releases) requires modeling of specific exposure scenarios; LCA is designed specifically to address systems without requiring unique location-specific information.

[46] Hendrickson, Lave, and Matthews (2006) notes the limitation of LCA outputs that are not linked to specific locations and exposures - "A typical [Life Cycle Inventory] of air pollution results in estimates of conventional, hazardous, toxic, and greenhouse gas emissions to the air. Even focused on this small subset of environmental effects, it is unclear how to make sense of the multiple outputs and further how to make a judgment as to tradeoffs or substitutions of pollutants among alternative designs."

[47] In the ecological realm, these kinds of translations are underdeveloped. The agency is aware of this ongoing limitation. For example, this conclusion has been drawn from several recent SAB reports, including EPA-SAB. 2003. "Underground Storage Tanks (UST) Cleanup & Resource Conservation & Recovery Act (RCRA) Subtitle C Program Benefits, Costs, & Impacts (BCI) Assessments: An SAB Advisory." (EPA-SAB-EC-ADV-03-001) and "Advisory on EPA's Superfund Benefits Analysis." (EPA-SAB-ADV-06-002). In addition, the SAB Committee on Valuing the Protection of Ecological Systems and Services is currently examining methods for addressing these limitations.

[48] The BEES model and supporting documentation are accessible at: http://www.bfrl.nist.gov/oae/software

[49] PaLATE does not allow for life cycle assessment of the inventory results, but in other life cycle models, impact assessment methods can be applied to inventory results to estimate environmental damages.

[50] BEES has reliable data for the use of fly ash in concrete, but it does not evaluate use of FGD gypsum in wallboard. SimaPro does allow evaluation of both fly ash and FGD gypsum but we prefer the U.S.-based data in BEES to conduct the fly ash analysis. For this reason, we do not use the same model for both analyses.

[51] The PaLATE model was used to evaluate beneficial use of fly ash in previous iterations of this chapter, but we omit the PaLATE analysis from this version in order to avoid comparisons of peer reviewed model findings to non-peer reviewed model findings for CCPs. It is important to note, however, that PaLATE relies on much of the same LCI data as the EIO-LCA model, which is presented in Appendix C.

[52] All concrete building product data in BEES (e.g., concrete columns, beams, walls, and slab on grade) use a 75-year lifespan assumption. Calculating unit impact values using data from any one of these products yields the same values. In addition to life cycle inventory data for concrete building products, BEES also includes data for a concrete parking lot pavement. The concrete parking lot data, however, use a 30-year lifespan assumption; calculating unit impact values using pavement data yields values that are approximately 2.5 times greater than the values calculated from building product data because the pavement data assume a 2.5 times shorter lifespan than building products.

[53] Fly ash replaces portland cement in concrete in a one to one ratio based on mass.

[54] Both the concrete beam made with and without blended cement assume a 60-mile round trip transport distance for portland cement and fly ash and a 50-mile round-trip transport distance for aggregate to the ready-mix concrete plant.

[55] We use the EcoInvent data set titled "Stucco, at plant/CH U" for this purpose.

[56] There may actually be emissions/dusting impacts associated with dewatering in a holding pond, but we have been unable to identify quantified estimates of these impacts. Alternatively, dewatering may be accomplished through mechanical processes but we were also unable to identify the energy impacts of mechanical dewatering.

[57] We do not model a transport differential between virgin and FGD gypsum to be consistent with the transport assumptions used in the BEES fly ash analysis, which helps preserve the comparability of the fly ash and FGD gypsum unit impact values. It is important to note, however, that an increasing number of new gypsum wallboard plants are being constructed adjacent to coal-fired power plants, so the transport distance of FGD gypsum to the wallboard manufacturing facility may, in some cases, be less than the transport distance of virgin gypsum.

[58] A discussion of general principals for allocation is presented in the International Standard on Environmental Management—Life Cycle Assessment— Goal and scope definition and inventory analysis (ISO 14041:1(E)), pp.11-12, and Annex B.

[59] It is important to stress that allocated impacts are not actual impacts associated with the beneficial use of the materials; in most cases use is significantly more beneficial than disposal. Instead, allocation is a means of placing the beneficial use materials in the context of their original production and recognizing that the processes that produce these byproducts may incur environmental costs.

[60] An example of a system in which dramatic changes in co-product value have driven production changes is the recent change in demand for ethanol, which has resulted in a significant increase in demand for agricultural by-products and has altered production decisions to meet this new demand.

[61] It is unclear from the documentation provided for BEES what impacts (e.g. virgin materials extraction, plant infrastructure, etc.) are modeled for portland cement production. For this reason, it is not possible to explain the differences in unit impact results between the FGD gypsum and fly ash analysis.

[62] Based on the assumption that an average residential customer uses 938 kilowatt-hours per month. Department of Energy, Energy Information Administration, "Energy Basics 101," http://www.eia.doe.gov/basics/energybasics101.html, accessed August 30, 2007.

[63] Based on 2000 USGS per capita water use estimate of 1,430 gallons per day. Lumia et al., United States Department of the Interior, United States Geological Survey, Summary of Water Use in the United States, 2000.

[64] As shown in Exhibit 4, a total of 49.6 million tons of CCPs were beneficially used in 2005.

[65] Based on the assumption that an average residential customer uses 938 kilowatt-hours per month. Department of Energy, Energy Information Administration, "Energy Basics 101," http://www.eia.doe.gov/basics/energybasics101.html, accessed August 30, 2007.

[66] Based on 2000 USGS per capita water use estimate of 1,430 gallons per day. Lumia et al., United States Department of the Interior, United States Geological Survey, Summary of Water Use in the United States, 2000.

[67] For example, the PaLATE model generates incremental effects on physical outputs arising from changes in roadway materials.

[68] Hendrickson, Lave, and Matthews (2006) ("A typical [Life Cycle Inventory] of air pollution results in estimates of conventional, hazardous, toxic, and greenhouse gas emissions to the air. Even focused on this small subset of environmental effects, it is unclear how to make sense of the multiple outputs and further how to make a judgment as to tradeoffs or substitutions of pollutants among alternative designs."), 29.

[69] Some in the LCA community refer to this as an LCA impact analysis, as opposed to the preceding LCA inventory analysis. Inventory analyses are those most commonly referred to as LCA. See Graedel and Allenby (1995).

[70] For an example of a full social cost & benefit analysis see Krupnick and Burtraw (1997).

[71] For example, this conclusion has been drawn from several recent SAB reports, including EPA-SAB. 2003. "Underground Storage Tanks (UST) Cleanup & Resource Conservation & Recovery Act (RCRA) Subtitle C Program Benefits, Costs, & Impacts (BCI) Assessments: An SAB Advisory." (EPA-SAB-EC-ADV-03-001) and "Advisory on EPA's Superfund Benefits Analysis." (EPA-SAB-ADV-06-002). In addition, the SAB Committee on Valuing the Protection of Ecological Systems and Services is currently examining methods for addressing these limitations.

[72] WARM can be accessed at http://yosemite.epa.gov/oar/globalwarming.nsf/WARM?openform. Version 8 of the model was used for this analysis. Available information indicate that Version 8 was last updated in August of 2006.

[73] WARM allows the user to define key modeling assumptions, such as landfill gas recovery practices and transport distance to MSW facilities. For landfill gas (LFG) control, we select the "National Average" setting, which calculates emissions based on the anticipated proportion of landfills with LFG control in 2000. For transport distances, we use the default setting (20 miles).

[74] It is important to note that the results reported by WARM for avoided greenhouse gas emissions and avoided energy use may not be directly comparable to those reported by the BEES model or PALATE model due to differences in the methodologies (including life cycle system boundaries) employed by each model.

[75] We use the EcoInvent data set titled "Stucco, at plant/CH U" for this purpose.

[76] There may actually be emissions/dusting impacts associated with dewatering in a holding pond, but we have been unable to identify quantified estimates of these impacts. Alternatively, dewatering may be accomplished through mechanical processes but we were also unable to identify the energy impacts of mechanical dewatering.

[77] We do not model a transport differential between virgin and FGD gypsum to be consistent with the transport assumptions used in the BEES fly ash analysis, which helps preserve the comparability of the fly ash and FGD gypsum unit impact values. It is important to note, however, that an increasing number of new gypsum wallboard plants are being constructed adjacent to coal-fired power plants, so transport distance of FGD gypsum to the wallboard manufacturing facility may, in some cases, be less than the transport distance of virgin gypsum.

[78] The tree is presented for the “single score” of all life cycle impacts, calculated using the Eco-Indicator 99-H, v2.04 impact assessment method.

[79] One limitation of substituting the Franklin U.S. electricity data set is that it represents low-voltage electricity production, but was used in place of medium-voltage European electricity production. This has the effect of slightly overstating the environmental impacts associated with electricity production in this analysis. Franklin U.S. electricity data set is the only data set in IEc’s version of SimaPro for average U.S. electricity production; a data set for medium-voltage U.S. electricity production is not available.

[80] Accessed at: http://www.eia.doe.gov/kids/energyfacts/science/energy

[81] DOE, "Emissions from Energy Consumption for Electricity Production and Useful Thermal Output at Combined-Heat-and-Power Plants," http://www.eia.doe.gov/cneaf/electricity last accessed Aug 29th 2007.

[82] DOE, "Carbon Dioxide Emissions from the Generation of Electric Power in the United States," http://www.eia.doe.gov/cneaf/electricity page/co2_report/co2report.html#electric, last accessed Aug 29th 2007.

[83] DOE, "Emissions from Energy Consumption for Electricity Production and Useful Thermal Output at Combined-Heat-and-Power Plants," http://www.eia.doe.gov/cneaf/electricity last accessed Aug 29th 2007.

[84] DOE, "Emissions from Energy Consumption for Electricity Production and Useful Thermal Output at Combined-Heat-and-Power Plants", http://www.eia.doe.gov/cneaf/electricity last accessed Aug 29th 2007.

In: Coal Combustion Waste: Management and Beneficial Uses ISBN: 978-1-61728-962-0
Editor: Daniel D. Lowell

Chapter 3

THE TENNESSEE VALLEY AUTHORITY'S KINGSTON ASH SLIDE: POTENTIAL WATER QUALITY IMPACTS OF COAL COMBUSTION WASTE STORAGE

Avner Vengosh

1. INTRODUCTION

On December 22, 2008, the retaining wall broke on a waste retention pond at the Tennessee Valley Authority (TVA) Kingston Fossil Plant, Tenn., and an estimated 4.1 million m^3 of coal ash slurry was spilled onto the land surface and into the adjacent Emory and Clinch Rivers (TVA, 2009). This was the largest coal ash spill in US history. The coal ash sludge spilled into tributaries that flow to the Emory River and directly into the Emory River itself (Figure 1), which joins to the Clinch River and flows to the Tennessee River, a major drinking water source for downstream users. With funds provided by the Dean of the Nicholas School of the Environment of Duke University, in January 2009 our team began a preliminary investigation of the potential environmental and health effects of the spill. This preliminary work (Vengosh et al., 2009; Ruhl et al., in revision) has thus far revealed three major effects: (1) The surficial release of coal ash formed a sub-aerial deposit that contains high levels of toxic elements (arsenic concentration of 75 mg/kg; mercury concentration of 150 &g/kg; and radioactivity (radium-226 + radium-228) of 8 pCi/g). These pose a potential health risk to local communities as a possible source of airborne re-suspended fine particles (<10 μm). (2) Leaching of the coal ash sludge in the aquatic environments resulted in severe water contamination (e.g. high arsenic content) in areas of restricted water exchange such as the Cove area, in a tributary of the Emory River. Further downstream, in the Emory and Clinch rivers, much lower levels of metals were found due to river dilution, but with metals concentrations above the background upstream levels. (3) High concentrations of mercury in downstream sediments of the Emory and Clinch rivers indicate physical transport of coal ash in the rivers. The high concentration of mercury and sulfate in the downstream river sediments could impact the aquatic ecosystems by formation of methylmercury in anaerobic river sediments.

A recent survey of the amount of coal ash generation in the United States revealed that 500 power plants nationwide generate approximately 130 million tons of coal ash each year, 43 percent of which is recycled into other materials. The remaining 70 million tons are stored in 194 landfills and 161 ponds in 47 states (Lombardi, 2009). An EPA study (USEPA, 2007) identified 63 coal ash landfills and ponds in 23 states where the coal sludge is associated with contaminating groundwater and the local ecosystem. One of the major potential hazards of coal ash storage in ponds is the continuous leaching of contaminants and their transport to the hydrological system. As such, the TVA coal ash spill provides a unique opportunity to evaluate the large-scale impact of coal ash leaching on the environment and water resources.

2. Fieldwork and Analytical Work

Coal ash sludge, sediments from the rivers, and water samples from tributaries, Emory and Clinch rivers, and springs near the spill area in Kingston and Harriman, Tenn., (Figure 1) were collected on two field trips on January 9-10 and February 6-7, 2009. Water sampling strictly followed the U.S. Geological Survey sampling protocol; trace metal and cation samples were filtered directly into new, high-purity acid-washed polyethylene bottles containing high-purity HNO_3 in the field for preservation using syringe-tip 0.45 μm filters. Trace metals in water were measured by inductively coupled plasma mass spectrometry (ICP-MS), mercury in sediments and coal ash by thermal decomposition, amalgamation, atomic absorption spectroscopy (Milestone DMA-80), and radium isotopes by gamma spectrometry at the Laboratory for Environmental Analysis of RadioNuclides (LEARN) at Duke University (http://www.nicholas.duke.edu/learn/).

3. Water Contamination

The chemical data show that surface water in the tributary that was dammed by the coal ash spill and turned into a standing pond ("the Cove" in the area of Swan Pond Circle Road; Figure 1) has relatively high levels of arsenic, calcium, magnesium, aluminum, strontium, manganese, lithium and boron. The concentration of arsenic was up to 86 μg/L in the Cove area (for reference, the EPA Maximum Contaminant Level in drinking water is 10 μg/L). The concentrations of these elements in springs that emerge into the Cove area are low, thus indicating that the shallow groundwater was not contaminated. We suggest that the non-contaminated groundwater discharges into the dammed tributary and causes leaching of metals from the sludge ash that was released from the TVA coal ash spill. Under restricted water exchange, the formation of standing water in the Cove resulted in contaminated surface water with high concentrations of arsenic, boron, strontium and other elements (Table 1).

Table 1. Concentrations of Trace Metals (µg/L) in Surface Water from the Cove Area (See Location in Figure 1)

Sample ID	Li	B	Al	Mn	Co	Ni	Cu	Zn	As	Se	Rb	Sr	Cr
The Cove													
TN1	13.24	431.89	-	846.9	2.11	3.98	1.54	16.54	69.59	2.44	15.57	578.4	1.88
TN1U	-	425.93	344.0	974.1	3.08	-	5.03	42.40	95.25	0.42	17.13	632.6	-
RC5	19.60	470.80	22	3014	6.96	8.97	1.57	47.16	85.56	3.75	23.76	1245	1.92
TN9	3.07	84.92	43.0	296.5	0.29	4.26	0.79	12.18	9.27	0.52	5.02	108.3	6.56
TN9U	-	112.89	197.0	331.8	1.15	-	3.48	24.86	12.60	-	6.25	120.1	-
RC8	7.39	229.63	40	556	1.89	1.67	2.77	36.64	20.70	1.83	6.35	456	0.47

U=unfiltered water

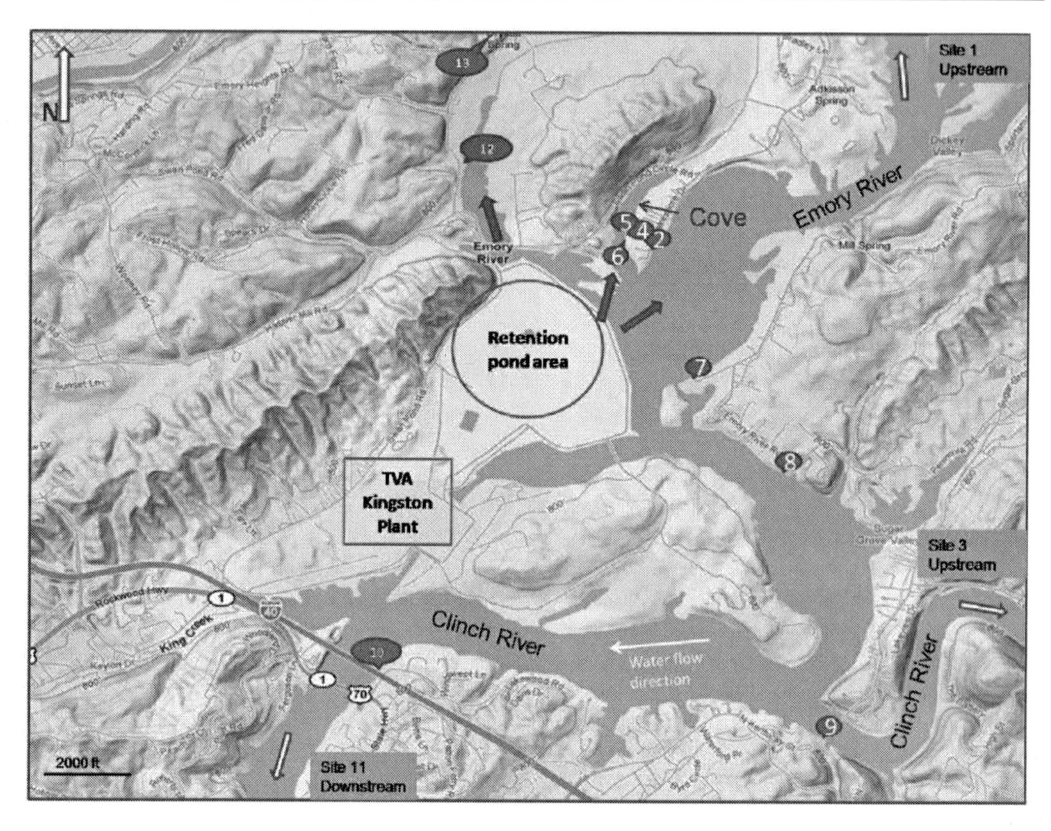

Figure 1. Map of the sampling sites of the TVA coal ash spill in Kingston, Tenn. From Ruhl et al. (in revision). Numbers refer to sampling sites in the vicinity of the TVA coal ash spill

In contrast, surface waters from the Emory River and Emory-Clinch River downstream from the breached dam show low concentrations of these metals, and all river inorganic dissolved constituents concentrations are below the EPA-Maximum Contaminant Levels. In spite of the absolute low concentrations, the metal contents in the downstream river samples are higher relative to the upstream river samples. For example, the arsenic levels in the downstream river samples are up to 3 μg/L relative to <0.4 μg/L in upstream rivers (Figures 2 and 3). We are able to detect these small changes due to the high sensitivity of our analytical instrument (inductively coupled plasma mass spectrometry; ICP-MS). This indicates that leaching of these metals from the coal ash in the river sediment is taking place in the river s, yet the massive dilution of the rivers reduces the content of these metals to below the MCL level. A report by TVA indicates that during storm events, remobilization of the coal ash resulted in short-term spikes of arsenic in the river (TVA, 2009). Remobilization of the river sediment by dredging could enhance metal leaching and contamination of the river water. Since dredging of the coal ash from the river bottom is an essential part of TVA restoration plan (TVA, 2009), it is essential to continue monitoring the water quality in order to evaluate the impact of dredging on the river water quality.

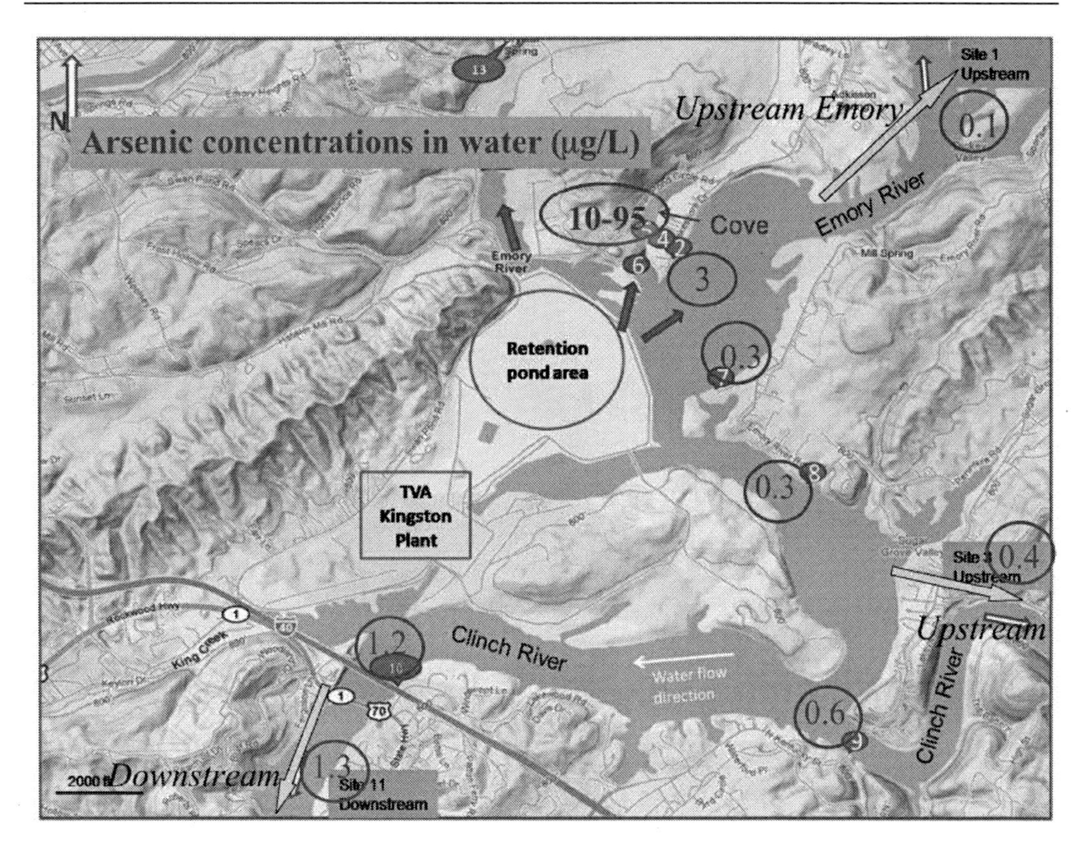

Figure 2. Map of the sampling sites of the TVA coal ash spill in Kingston, Tenn., with concentrations of arsenic (µg/L) in surface waters associated with the TVA coal ash spill

4. River Sediment Contamination

The mercury concentration in the TVA coal ash sludge (an average of 151.3±15.9 µg/kg) is higher than background soil in Tennessee (45-50 µg/kg) (USGS survey data, 2004). These concentrations are consistent with mercury concentrations previously reported in fly ash (100 to 1500 µg/kg) (Sanchez et al., 2006). In the sediments of the Emory and Clinch rivers, the mercury concen tration increases from 29.7-43.3 µg/kg in upstream sediments, to 115-130 &g/kg in downstream sediments from the spill site (Figure 4). The mercury concentrations of the upstream sediments are consistent with previously reported Hg data for the lower Clinch River and for the overall Tennessee River (USGS survey data, 2004). However, the relatively high mercury concentrations in the downstream river sediments could indicate a significant transport of the coal ash in the river and deposition in the river sediments. We measured relatively high mercury in sediments at Site 10, downstream from the underwater bar that was built to prevent migration of the ash (Figure 1). A simple mass balance between the mercury content of coal ash (150 µg/kg) and background soil (50 µg/kg) suggests that the downstream river sediment at Site 10 was composed of about 66 percent ash. The assumption that mercury in the river sediments is derived from only redistribution of coal ash needs to be confirmed by further research.

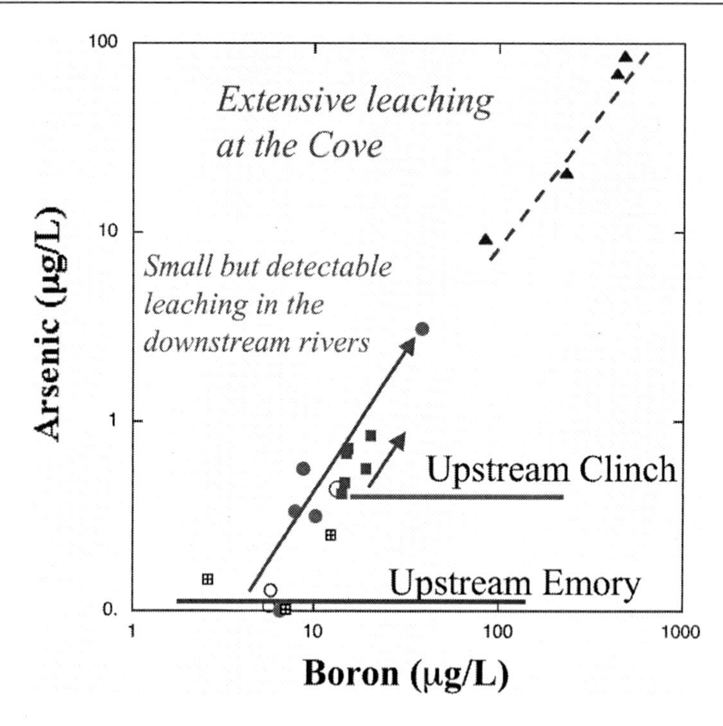

Figure 3. Concentrations of arsenic versus boron in Emory (blue) and Clinch (green) rivers and in groundwater (squares) in logarithmic scale. Note the relative enrichment of the downstream Emory and Clinch river samples relative to the respective upstream river samples.

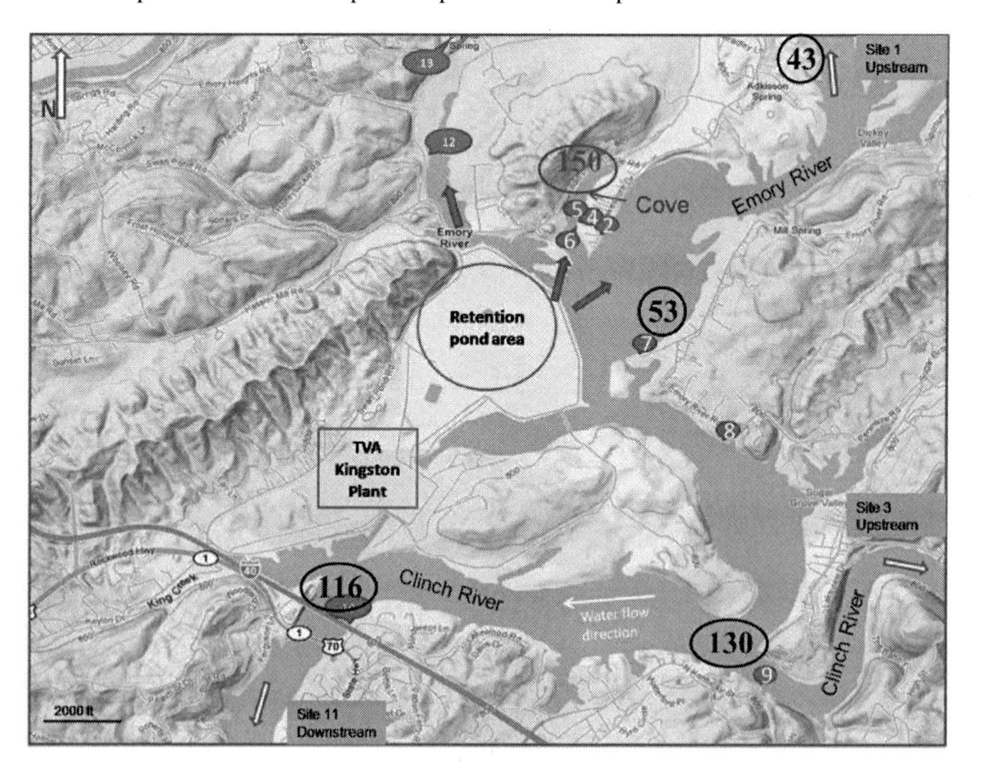

Figure 4. Map of the sampling sites of the TVA coal ash spill in Kingston, Tenn., with concentrations of mercury (μg/kg) in sediments (black) and coal ash (red) Data from Ruhl et al. (in revision)

The ecological impact of high mercur y (and arsenic) in the river sediments has not been determined as yet. We hypothesize that accumulation of coal ash in the river sediments might generate transformation of elemental mercury to methylmercury by anaerobic bacteria in river sediments. Forming of methylmercury in river sediments is a concern because of bioaccumulation of methylmercury in food webs. In addition, accumulation of As-rich fly ash in bottom sediment and leaching of arsenic to pore water might cause fish poisoning via both food chains and decrease of benthic fauna that is a vital food source. These potential hazards should be monitored.

5. Conclusions

- Leaching of the coal ash sludge in the aquatic environments resulted in severe water contamination (e.g. high arsenic content) in areas of restricted water exchange - the Cove area.
- Further downstream in the Emory and Clinch rivers, much lower levels of these metals were found due to river dilution, but with metal concentrations above the background upstream levels.
- Remobilization of the river sediment by dredging could enhance metal leaching thus it is essential to continue monitoring the water quality in order to evaluate the impact of dredging on river water quality.
- High concentrations of mercury in downstream sediments of the Emory and Clinch rivers suggest physical transport of coal ash in the rivers.
- The high concentration of mercury in the downstream river sediments could impact the aquatic ecosystems by formation of methylmercury in anaerobic river sediments. Forming of methylmercury in river sediments is a concern because of bioaccumulation of methylmercury in food webs.
- Accumulation of As-rich ash in bottom sediment and leaching of arsenic to pore water might cause fish poisoning via both food chains and decrease of benthic fauna that is a vital food source.

References

Lombardi, K. (2009). *Coal Ash: The Hidden Story*, The Center for Public Integrity (http://www.publicintegrity.org/articles/entry/1 144/).

Ruhl, L., Vengosh, A., Dwyer, G. S., Hsu-Kim, H., Deonarine, A., Bergin, M. & Kravchenko, J., (in revision). A Preliminary Investigation of the Environmental and Health Effects of the Coal Ash Spill at Kingston, Tennessee. *Environ. Sci. Technol.* (in revision).

Sanchez, F., Keeney, R., Kosson, D. & Delapp, R. (2006). *Characterization of Mercury-enriched Coal Combustion Residues from Electric Utilities Using Enhanced Sorbents for Mercury Control*; U.S. EPA: Washington, D.C .

Tennessee Valley Authority (2009). *Corrective Action Plan for the TVA Kingston Fossil Plant Ash Release*. (http://www.tva.gov/kingston/index.htm).

U.S. Geological Survey. (2004). The National Geochemical Database; *U.S. Geological Survey Open-File Report 2004-1001*; Reston, VA.

U.S. Environmental Protection Agency. (2007). *Coal Combustion Waste Damage Case Assessments*. U.S. EPA, Office of Solid Waste, July 9, 2007.

Vengosh, A., Ruhl, L. & Dwyer, G. S. (2009). *Possible Environmental Effects of the Coal Ash Spill at Kingston*, Tennessee. Phase I: Preliminary Results. Internal Report; Nicholas School of Environment, Duke University.

In: Coal Combustion Waste: Management and Beneficial Uses ISBN: 978-1-61728-962-0
Editor: Daniel D. Lowell

Chapter 4

TESTIMONY OF TOM KILGORE, PRESIDENT AND CHIEF EXECUTIVE OFFICER, BEFORE THE SUBCOMMITTEE ON WATER RESOURCES AND THE ENVIRONMENT, HEARING ON "TENNESSEE VALLEY AUTHORITY"

OPENING STATEMENT

Chairwoman Johnson, Ranking Member Boozman, and members of the Committee. I appreciate this opportunity to discuss the coal ash spill at the Tennessee Valley Authority's (TVA) Kingston Fossil Plant, the actions taken in response to the event, and our progress and plans for remediation of the site and protection of the environment.

The incident being discussed today occurred at TVA's Kingston Fossil Plant in Roane County, Tennessee. On behalf of TVA, we deeply regret the failure of the ash storage facility dike, the damage to adjacent private property in the Swan Pond community, and the impact on the environment. We are extremely grateful that no one was seriously injured.

TVA is committed to cleaning up the spill, protecting the public health and safety, and restoring the area. In the process, we will look for opportunities, in concert with the leaders and people of Roane County, to make the area better than it was before the spill occurred. This commitment will stand because TVA is part of the Kingston community through our employees who live and work there, and through the partnership of our historic mission to work for the economic progress of the Tennessee Valley region.

We are also committed to sharing information and lessons-learned from this event and the recovery with those in regulatory and oversight roles, such as this committee, and with others in the utility industry.

Today marks the 99th day since the spill occurred. We have made steady progress in the initial recovery work, including development of a Corrective Action Plan that includes comprehensive monitoring of the air, water and soil. It is important to note that according to the Tennessee Department of Health, the environmental monitoring analyzed to date has not shown any adverse health threat to the immediate or surrounding community, including air quality or drinking water supplies. On March 19, we began the initial phase of dredging ash from the Emory River channel adjacent to the failed storage facility. This activity is being thoroughly monitored and precautions are in place to prevent or minimize environmental impacts during the dredging process. The dredging plan was approved by the Tennessee Department of Environment and Conservation (TDEC) and the U.S. Environmental Protection Agency (EPA).

An investigation by an outside engineering firm is under way to determine the root cause of the event. The results of the report are expected this summer. In the meantime, we are proceeding with the recovery work. We understand this is a difficult time for residents of the Kingston community, and we are working to make things right.

Our objectives are:

(1) To protect the health and safety of the public and recovery personnel.
(2) Protect and restore environmentally sensitive areas.
(3) Keep the public and stakeholders informed and involved in formulation of the response activities.
(4) Clean up the spill and improve the area wherever possible in coordination with the people of Roane County.

My comments today will cover three areas: what occurred; the response and initial recovery thus far; and TVA's plans going forward for full recovery and site remediation. Before discussing the Kingston event, I want to briefly describe TVA and its mission.

About TVA

TVA is a corporate agency and instrumentality of the United States government, is wholly owned by the United States, and is the nation's largest public power supplier. Under the TVA Act, TVA's hydroelectric dams and other power generation facilities are designed and operated as part of a multipurpose system to help improve navigation, control floods, meet national defense needs and promote the development of the Tennessee Valley region. Since 1959, in accordance with the direction of Congress, TVA has operated the power system to be financially self-supporting. Today, we use our power revenues to buy fuel, pay wages, service our debt, maintain assets, and fund our environmental stewardship and economic development activities.

In partnership with 158 local utilities, TVA provides reliable, affordable electricity to nine million people and 650,000 businesses in Tennessee and parts of six surrounding states. The 158 local utilities are our wholesale customers. The local utilities purchase TVA power for retail sale to their residential, commercial and industrial customers. TVA also sells power directly to about 60 large industrial customers and federal installations, such as Oak Ridge National Laboratory.

TVA has stewardship responsibilities for the Tennessee Valley region's natural resources, including the nation's fifth-largest river system. TVA's management of an integrated river systems and innovative watershed management are recognized as national and international models for government and community collaboration for improving and protecting water quality. TVA also is a catalyst for economic development and job creation throughout its 80,000-square-mile service area, working in partnership with local governments and economic development agencies.

KINGSTON FOSSIL PLANT AND FLY ASH STORAGE

The ash spill that is the subject of today's hearing occurred at Kingston Fossil Plant, which is about 40 miles west of Knoxville, Tennessee. Construction began on Kingston in 1951 and it was completed in 1955. The plant was built in accordance with congressional authorization, primarily to meet the defense needs of the nation – specifically, to provide power for the production of atomic defense materials at Oak Ridge, Tennessee.

Today, Kingston is part of a diverse mix of generating resources that TVA uses to supply electricity for nine million people in our service region in the Southeast. About half of our nation's electricity supply comes from coal, and TVA's supply is similar. While we are working to increase the amount of carbon-free generation, about 60 percent of TVA electricity comes from coal. And like utilities nationwide, we must manage the ash that is a by-product of coal-fired power production.

At the Kingston plant, ash material that remains after the coal is burned is stored in a wet ash pond. Six of TVA's eleven fossil plants use wet fly ash storage cells. The other five plants use a dry fly ash storage method. All of TVA's ash disposal sites are engineered facilities governed by the permit requirements of the states where they are located. The storage cells are surrounded by dikes, and the facilities have engineered drainage systems and water runoff controls.

The storage areas at all TVA fossil plants undergo a formal inspection annually, and other inspections are conducted on a daily and quarterly basis. The storage cells at Kingston are visually checked daily by plant personnel. In addition, plant personnel inspect for seepage on a quarterly basis. Annually, TVA engineering staff members perform a comprehensive inspection and document their findings and recommendations. Kingston's most recent inspection was in October 2008, and the formal report was being compiled at the time of the event. The completed report is now posted on the TVA Web site. Nothing that would indicate a catastrophic failure was likely to occur was observed during the annual inspection.

HISTORY OF THE EVENT AND EMERGENCY RESPONSE

On Monday, December 22, 2008, between midnight and 1 a.m., a portion of the dike on the northwestern side of the Kingston storage cell failed, releasing about 5.4 million cubic yards of fly ash and bottom ash onto land and adjacent waterways, including the Emory River, which flows into the Clinch River near the plant. The Clinch then flows into the Tennessee River. The released ash covered about 300 acres of which eight acres were privately-owned lands, not owned or managed by TVA. TVA has now purchased all but one of those acres. The spilled material covered most of the Swan Pond Embayment and reservoir shorelines, along with parts of Swan Pond Road and Swan Pond Circle and portions of the rail line used for coal deliveries to the Kingston plant. Surveys done since the event show that ash was released from about 60 acres of the 84-acre storage facility, which is surrounded by dikes about 60-feet high.

I received a call notifying me about the failure shortly after 1 a.m. and arrived at the plant within the hour. The initial response by the Roane County Office of Emergency Management and Homeland Security personnel, along with the Tennessee Emergency Management Agency, was excellent; and we will always be grateful for their swift and professional response. Other agencies were notified, including the National Response Center.

Our first concern was for the safety of the neighbors near the plant. With the help of the Roane County response personnel, we learned about 5 a.m. that there was no loss of life and no injuries that required medical attention. We ordered visual inspections of the ash retention dikes at all of our other plants to detect any changes in conditions, and those inspections continue on a daily basis.

Our first priority was to help the people immediately impacted, especially the three families whose homes were severely damaged and deemed uninhabitable. We ensured they were safe and that they had temporary housing, meals, and other necessities. We established a team of TVA employees and retirees to provide a single point of contact for each family impacted to ensure their needs were met and concerns addressed.

We set up a 1-800 number and opened a Community Outreach Center in Kingston that was open initially seven days a week to handle property damage claims and respond to residents' questions and concerns. Claims adjustors and field staff were provided by a national claims management company at the outreach center to conduct on-site damage assessments, and TVA Police supported local law enforcement in maintaining security for homes in the affected area. The Community Outreach Center is now open from 2 to 6 p.m. Monday through Friday. The center has been in touch with almost 750 households and received nearly 400 real estate-related claims and 241 health-related concerns.

In the early stages of the event, TVA followed its approved Agency Emergency Response Plan which provides an agency-wide response to emergencies or threats that require integrated agency action. The Senior Management Executive was responsible for directing the emergency response through the Agency Coordination Center. The U.S. Environmental Protection Agency (EPA) joined TVA, TDEC, and other agencies in a coordinated response and provided oversight and technical advice for the environmental response portion of TVA's activities. TVA transitioned its emergency response to a Unified Command Center as defined by the National Incident Management System. On January 11, EPA turned the lead federal role over to TVA, and the Unified Command structure was transitioned into an onsite recovery response organization, using TVA's Fossil Emergency Plan procedure (FPG.EP.14.000).

Initial results of all environmental sampling and updates on the response activities were communicated to the public through media briefings at the Joint Information Center that was established at the Roane County Rescue Squad headquarters building near Kingston. Other information and test results are posted on the TVA public Web site.

In addition to media briefings at the Joint Information Center, TVA hosted a public open-house with representatives from key state and federal agencies on January 15 at Roane State Community College where residents could pose questions to experts and obtain information. The latest open house was held last night (March 30) at the community college to bring residents up to date and answer their questions. TVA representatives attended several public meetings and other forums to provide information and answer questions. Information was made available in the form of Material Safety Data Sheets to help make residents aware of potential hazards and actions they could take to minimize any risk.

Within the first month, TVA began purchasing affected properties using appraisals by state certified residential and general appraisers. Offers were made based on the higher of two independent appraisals. The appraisals are based on property values on December 20, 2008, before the spill. In addition, an amount significantly above the fair market value is added to the appraised value to assist the property owner in reestablishing residence. Property owners who accept the offers also are given first right of refusal to re-purchase the property at market value if TVA decides to sell the property in the future. TVA has extended offers on 92 tracts in the area, including primary and secondary residences, vacant lots, and two businesses.

ENVIRONMENTAL EFFORTS

A principal concern regarding air quality comes from airborne particulates in the form of dust blown from dry ash deposits that can irritate the respiratory system if breathed over long periods. We took immediate measures to keep the ash residue damp and monitor air quality in the area. The dust suppression measures were expanded during the first week to include aerial grass seeding and mulching with straw to provide a vegetative cover to minimize dust and erosion. The seeding measures covered about 213 acres. We also are conducting a continuous schedule of watering from pumping trucks and employing vacuum sweeper trucks on paved roads in the area. Three wheel-washing stations are installed for heavy trucks leaving the site to prevent the spread of ash onto roads. TVA has prepared and implemented plans for air monitoring and dust suppression activities. These TVA plans were developed with regulatory oversight by TDEC and EPA. The dust suppression plan is being updated to reflect additional suppression techniques. Both agencies have visited the site to monitor TVA's progress in implementing the plans.

The air monitoring results are a measure of the efficacy of dust suppression efforts. Air monitoring results to date indicate airborne particulate levels (PM 10 and PM2.5) within daily National Ambient Air Quality Standards. Metals analysis of the airborne dust indicates levels in the range of normal background levels and not at a level of a health concern. TVA installed new PM2.5 air monitors (previous PM2.5 monitors were demobilized on February 3) and placed them into service on February 12. Air monitoring is done 24 hours a day at fixed stations located in residential areas near the plant and on the plant site.

Testing of offsite soil samples shows that metals are well below the limits for classification as a hazardous waste. They are 10 to 100 times below the limits for toxic metals. The trace concentrations of metals in the offsite material sampled are consistent with and generally lower than that of the historic sampling results from the storage cell. The data shows that the concentrations of most metals in the deposited ash are not significantly different from concentrations found in natural, non-agricultural soils in Tennessee, with the exception of arsenic. Total arsenic results were above the average that occurs naturally, but well below levels found in soils that are well-fertilized and significantly below the limits to be classified as a hazardous waste.

According to the Tennessee Department of Health, public drinking water supplies continue to meet state and federal drinking water standards, and private wells and springs tested within four miles of the site are not impacted by the coal ash release. TVA will continue to work with TDEC to monitor the water quality at private wells and springs in the vicinity of the ash release to ensure their protection. Periodic monitoring of private wells and springs located within approximately 0.25 mile of ash-impacted property bordering the Emory River and its tributaries will be performed. Some 47 land parcels having inferred well or spring water supplies are indicated within the designated monitoring region.

Early-warning groundwater monitoring wells are being installed, as needed, at selected locations to ensure protection of water supplies deemed by TDEC to be at potential risk. Sampling frequency will vary from quarterly to semiannually during the first year depending on proximity of each well or spring to ash deposits. The frequency and ultimate duration of sampling of off-site wells and springs will be re-evaluated annually by TVA and TDEC based on monitoring results and perceived risks. Water samples will be analyzed for several constituents including radio-nuclides.

Air, water and soil sampling by TVA and TDEC includes: more than 27,000 air samples; more than 1,050 utility and surface water samples; more than 100 well and spring water samples taken from within a four-mile radius of the spill site; 81 ash samples; and 47 soil and sediment samples. The City of Kingston has also conducted more than 140 tests on utility drinking water. Each agency uses certified laboratories for testing. Sampling results have not indicated a heath concern, according to the Tennessee Department of Health.

I know that technical data and monitoring equipment do not make the physical effects of the situation go away. But I hope that the results of the environmental monitoring data during the past three months and the objectivity provided by multiple agencies and certified labs will help reassure the public. The information is available on the TVA Web site, along with other information, including the Corrective Action Plan.

TVA is developing a plan to respond to individual health concerns, including a process for determining whether there are health effects that may be related to the ash released from Kingston. We are in the process of contracting with Oak Ridge Associated Universities (ORAU) to provide community members and the local medical community with access to medical and toxicology experts who have experience and knowledge in the health effects related to the contaminants in the Kingston ash. ORAU has expertise in public health communication, design of medical monitoring programs, and independent verification of the clean-up of contaminated sites. ORAU is a consortium of 100 academic universities that collaborate to advance scientific research and education.

Recovery Actions

In addition to ensuring the health and safety of the public and our employees, TVA then moved quickly to stabilize, contain, and plan for recovery of the ash material. In response to an order from the Tennessee Commissioner of Environment and Conservation, TVA prepared a Corrective Action Plan that was submitted to the State of Tennessee and the EPA.

The recovery work began with clearing more than 350,000 cubic yards of material from the areas around Swan Pond Road, Swan Pond Circle, the rail line, and nearby sloughs. The two roads are now open for use by construction vehicles involved in the recovery, and 2,100 feet of rail line was reconstructed and returned to service for coal deliveries to the plant.

Reconstruction of the rail line within a month of the event avoided the potential use of local roads for coal deliveries and assured efficient use of the Kingston plant power output for the region's electricity supply.

About 5,800 feet of drainage trench has been installed in the Swan Pond Embayment, and 6,400 feet of drainage trench is installed around the roads and rail line. In addition, 11,000 feet of isolation barrier was installed in the affected areas to contain the ash.

To prevent migration of the ash from the Swan Pond Embayment and the Emory River channel a 615-foot-long underwater rock weir was constructed across a section of the Emory River, and a dike was constructed along the embayment.

TVA is also managing the flows of the Clinch and Tennessee Rivers in the Kingston area to minimize downstream movement of the ash and to maintain a positive flow downstream to protect the integrity of the Kingston water supply intake. The water intake is on the Tennessee River about one-half mile upstream from the confluence of the Clinch and Tennessee Rivers.

The Corrective Action Plan submitted under order to the Tennessee Department of Environment and Conservation and to EPA provides a framework for making future decisions about environment remediation, monitoring during cleanup activities, for protecting water supplies, protecting work and public health, and management of spilled ash and future ash produced at Kingston.

The plan proposes the formation of an Interagency Team consisting of personnel from all involved and interested federal, state and local agencies. We propose that the team be involved in all steps of the cleanup and recovery effort. We also plan to develop a Community Involvement Plan to provide a structure for public review and input into the recovery and remediation.

Recovery Effort Milestone - Dredging

The first major phase of the recovery was the start of dredging operations on March 19 in the Emory River channel adjacent to the failed storage cell. Construction of the dike and weir support the first phase of dredging, which serves as a pilot for future dredging operations. A plan for the first phase was developed for TVA by an environmental services contractor and has been approved by the state and the EPA. The plan is designed to remove an estimated 2 million cubic yards of ash material.

An Environmental Assessment, consistent with the National Environmental Policy Act, was developed for the dredging operations, and a comprehensive environmental sampling plan was submitted for review to state and federal regulators. The sampling plans include six floating hydro-labs to monitor key environmental criteria, such as dissolved oxygen and turbidity, during the dredging operations. Containment booms are also being installed on the water to prevent migration of any floating ash material.

Ash is being dredged from the Emory River channel, de-watered, and temporarily stored at a prepared site on the plant property until an approved process is in place for longterm disposal or storage. The de-watering area is sloped to drain into the plant's existing ash pond and drainage has been engineered at the site to contain the runoff. Groundwater wells have been drilled in the area for monitoring.

Plans call for dredging only to a depth that will restore flow to the original channel without disturbing existing "legacy" and native river sediments. Restoring original flow to the channel will lessen the possibility of flooding upstream on the Emory River in the event of unusually heavy rains. We have advised residents in the potential flooding areas about the situation and have assured them that TVA will assume responsibility for any damage to homes above the traditional flood stage.

While most of the fly ash deposited in the water sank, there was a lighter, inert part of the ash that floats. This hollow, sand-like material, called cenospheres, is collected and sold for use in a variety of products, including cosmetics and bowling balls. We have used more than 12,000 feet of boom skimmers to collect and dispose of more than 3.2 million gallons of slurry containing this material. The containment booms and other equipment will be used to collect this material released during dredging.

At this time, future plans call for proposing two more phases of dredging. The second phase would restore the river channel to its original depths, and the third phase would focus on removing ash deposits that are outside of the Emory River channel.

Going Forward

TVA has commissioned a comprehensive study of all its coal by-product storage facilities by an outside engineering firm. The study includes invasive testing of dike walls to evaluate their composition and structural integrity. We are also looking at the feasibility of converting to dry fly ash storage at all six of our plants where wet storage is used.

TVA has committed to ceasing wet ash storage in the failed cell at Kingston, and the cell must be closed and capped. This will be done once the conclusions of the root cause analysis are known and the site subsurface investigations are complete. In early January 2009, TVA retained a global engineering firm that possesses substantial experience in design, construction quality management, and forensic failure analyses of dikes, containment ponds, and landfills, to conduct an independent Root Cause Failure Analysis (RCA) of the Kingston dike failure.

Data from both the Root Cause Failure Analysis and the impoundment assessments are shared with TDEC, EPA and TVA's Office of Inspector General, who comprise a Structural Integrity Team.

We do not have a completed cost schedule for the recovery and remediation, but based on the dredging and other identified tasks ahead, we estimate that it will cost between $525 million to $825 million (not including litigation, penalties or settlements) depending on methods of disposal and other variables. We are evaluating several potential sources for funding the recovery. These include insurance, using a portion of a trust fund established for the retirement of non-nuclear assets, using debt for funding over a longer period, and recovering some of the costs through rates.

Widows Creek

The committee staff requested that I also provide information about the accidental spill of slurry from the Gypsum storage pond at TVA's Widows Creek Fossil Plant near Stevenson, Alabama, that occurred on January 9, 2009. The spill occurred when a cap dislodged on a 36-inch diameter drainage pipe that was no longer in use due to reconfiguring of the storage pond over the years.

The event allowed water from the gypsum pond to drain into an adjacent settling pond, filling it to capacity and causing it to overflow. Although most of the overflow was contained in the settling pond, some did drain into adjacent Widows Creek and into a slough on the Tennessee River. The event was discovered about 6 a.m. by plant workers who were conducting a routine inspection of the ponds.

The impoundment contains byproducts from the scrubbers that clean sulfur dioxide from the plant's coal-burning emissions. Scrubbers produce a number of byproducts while cleaning the air, the primary one being calcium sulfate - commonly known as gypsum. Beneficial uses of gypsum are numerous and include drywall and cement manufacturing. Gypsum is also used as a soil amendment in place of lime in agricultural and construction activities.

We notified appropriate federal, state and local authorities, and water sampling was conducted that indicated there was no danger to water supplies in the area or downstream. TVA, the EPA and the Alabama Department of Environmental Management estimate that less than 5,000 cubic yards of material entered the waters.

A cleanup operation was begun, and repairs and improvements were made to the storage ponds, including pouring concrete into the abandoned drain pipe. An investigation showed that a major contributing factor was omission of the abandoned drainpipe on engineering drawings of the storage pond.

Continuing Commitment

As I stated earlier, TVA is an integral part of the Roane County community. About 300 TVA employees live and work in the area, and they care deeply about their community. We will continue to reach out to Roane County residents to keep them informed and ensure they have the information they need. We will continue working, as well, with federal, state, and local elected officials and agencies, and with you and other members of Congress.

We are committed to do a first-rate job of remediation of the problems caused by the spill and ensure the integrity of all of our coal by-product storage facilities across the TVA system. Thank you for the opportunity to discuss our recovery efforts.

I look forward to your questions.

In: Coal Combustion Waste: Management and Beneficial Uses ISBN: 978-1-61728-962-0
Editor: Daniel D. Lowell

Chapter 5

TESTIMONY OF STAN MEIBURG, ACTING REGIONAL ADMINISTRATOR, REGION 4, U.S. ENVIRONMENTAL PROTECTION AGENCY, BEFORE THE SUBCOMMITTEE ON WATER RESOURCES AND THE ENVIRONMENT, HEARING ON "TENNESSEE VALLEY AUTHORITY"

Madam Chairwoman and members of the Subcommittee, thank you for the opportunity to provide testimony on the U.S. Environmental Protection Agency's (EPA's) role in the response and clean up of the release of coal ash from the Tennessee Valley Authority (TVA) Kingston Fossil Plant in Harriman, Roane County, Tennessee. I will discuss the actions EPA has taken as part of the response to this release, as well as our current and planned actions to ensure that the ash removal and disposal is conducted in a manner that protects public health and the environment.

RESPONSE TO KINGSTON COAL ASH RELEASE

On December 22, 2008, at 1:00 a.m., an ash disposal cell at the TVA Kingston Fossil Plant failed, causing the release of an estimated 5.4 million cubic yards of fly ash to the Emory and Clinch Rivers and surrounding areas. The release extended over approximately 300 acres outside the ash storage area. The failed cell was one of three cells at the facility used for settling the fly ash. The initial release of material created a wave of water and ash that destroyed three homes, disrupted electrical power, ruptured a natural gas line in a neighborhood located adjacent to the plant, covered a railway and roadways in the area, and necessitated the evacuation of a nearby neighborhood.

Shortly after learning of the release, EPA deployed an On-Scene Coordinator to the site of the TVA Kingston Fossil Plant coal ash release. EPA joined TVA, the Tennessee Department of Environment and Conservation (TDEC), the Roane County Emergency Management Agency, and the Tennessee Emergency Management Agency (TEMA) in a coordinated response (i.e., unified command in the National Incident Management System). EPA provided oversight, as well as technical advice, for the environmental response portion of TVA's activities. TVA has conducted extensive environmental sampling and shared results with EPA personnel. As discussed in more detail below, EPA staff and contractors have also conducted extensive independent sampling and monitoring to evaluate public health and environmental threats. In addition to providing information on environmental conditions at the site, EPA's data have also served as an independent verification of the validity of the TVA data.

EPA sampling has included: surface waters of the Clinch and Emory Rivers, municipal water supply intakes, finished water (distributed from the water treatment plant) from potentially impacted public water systems, soils, private drinking water wells, and coal ash. EPA also monitored airborne particulate levels in areas of ash deposition. The multimedia data are being used to determine appropriate response measures that are protective of the environment and human health.

In the aftermath of the incident, EPA sampled the coal ash and residential soil to determine if the release posed an immediate threat to human health. Sampling results for coal ash contaminated residential soil showed arsenic, cobalt, iron, and thallium levels above the residential Superfund soil screening values. Sampling results also showed average arsenic levels in the Kingston coal ash and coal ash contaminated residential soil above the EPA Region 4 Residential Removal Action Levels (RALs). RALs are used to trigger time-critical removal actions while soil screening values, are used as a point of departure for EPA to take any action to investigate and/or remediate a release. In response to exceedances of RALs for ash contaminated residential soils, TVA relocated residents to interrupt this soil exposure pathway. All other compounds in the ash and ash contaminated residential soil were below these soil screening values. EPA also analyzed the coal ash under the Toxicity Characteristic Leaching Procedure to determine whether the material would be classified as a hazardous waste, were it not for an exemption under the Resource Conservation and Recovery Act. The analysis showed the material would not be classified as a hazardous waste.

Since the failure, EPA, TDEC, and TVA have sampled multiple locations along the Clinch and Emory Rivers. Those sampling efforts detected heavy metals known to be contained in coal ash, but concentrations were below applicable limits. To date, almost 800 surface water samples have been taken by TVA and TDEC, ranging from two miles upstream of the release on the Emory River to approximately eight miles downstream on the Clinch River. Sampling results of untreated river water showed that some metals were elevated just after the incident, including arsenic, cadmium, chromium, and lead. Elevated levels of metals in untreated river water were observed again after a heavy rainfall on January 6, 2009. However, subsequent sampling events found decreasing amounts of suspended ash, and showed metals concentrations below drinking water limits.

For drinking water, concentrations measured on December 23, 2008, near the intake of the Kingston Water Treatment Plant (WTP) were below federal Maximum Contaminant Levels (MCLs) for drinking water with the exception of elevated thallium levels. Subsequent EPA testing on December 30, 2008, of samples at the same intake found that concentration levels for thallium had fallen below the MCL. On December 29, 2008, and again during the December 30, 2008, sampling event, EPA sampled the finished water at the Kingston WTP. These samples were below MCLs. Additional testing conducted during the December 30, 2008, sampling event confirmed that samples from the Cumberland and Rockwood WTPs did not exceed MCLs. A regular sampling program implemented by TDEC at the Kingston WTP is in place and continues in operation.

Some residents near the site rely on private wells as their source of drinking water. EPA identified and sampled several potentially impacted residential wells in the immediate area on December 30, 2008. No contaminants above MCLs were detected. In coordination with EPA testing, TDEC offered to sample all residential wells within a four-mile radius of the facility. As of March 26, 2008, TDEC has taken 112 water samples (both spring water and well water). To date, all of the samples have met the Drinking Water MCLs. Well sampling is a voluntary process that must be initiated by each resident, and TDEC continues to receive and accommodate sampling requests within four miles of the facility.

EPA and TDEC recognize that windblown ash poses a potential risk to public health. With EPA oversight, TVA commenced air monitoring for coarse (10 microns in size) and fine (2.5 microns in size) particulate matter (PM_{10} and $PM_{2.5}$, respectively). Concurrently, EPA and TDEC commenced monitoring for PM_{10} and $PM_{2.5}$ to validate TVA's findings. To date, almost 26,000 air samples have been collected. Particulate levels in the air have measured below the National Ambient Air Quality Standards for these parameters. TVA has constructed five air monitoring stations in residential neighborhoods surrounding the site and developed a strategy for air monitoring throughout the duration of the clean up. TVA is also implementing a number of dust control measures, including water trucks, vehicle cleaning, and erosion control mulch.

TVA also obtained several air samples on TVA property to measure potential levels of specific contaminants of concern in the air. No constituents were detected with the exception of silica in a single sample. After consultation with the Agency for Toxic Substances and Disease Registry (ATSDR), the level of silica detected was determined not to pose an imminent threat to public health. Sampling results for sediment, air, and water testing are available on the TDEC, TVA, and EPA Region 4 websites.

While protection of public health and safety was the primary concern during the initial phase of emergency response, EPA's mission also calls for protection of the environment, in this case the long-term ecological health of the Emory and Clinch Rivers. As part of its response, TVA constructed an initial rock weir across the Emory River to minimize downstream sediment transport, and a second weir to contain ash which is located in Swan Pond Embayment adjacent to the Emory River. A detailed ecological assessment will determine appropriate future actions to restore the functions of this aquatic system and its tributaries. TVA has also constructed drainage channels across the ash in the Swan Pond Embayment to reduce the potential for flooding in the three tributary systems that feed the embayment and to reduce water flowing through the ash. TVA has submitted a storm water construction permit for the embayment area, and this permit has been approved by TDEC. This permit involves the construction of two additional dikes at the upstream extent of the ash in the tributaries to reduce the mixing of stormwater flows with the ash, and a stormwater pond for treatment. The pond is presently being constructed adjacent to the second weir across Swan Pond Embayment.

KEY CLEANUP ACTIVITIES

The ash disposal cell which failed had been permitted by TDEC as a Class II Solid Waste Landfill under State regulations, and TDEC remains the lead oversight agency for this clean up. On January 12, 2009, the Commissioner of TDEC issued an order to TVA that among other things required TVA to submit a Corrective Action Plan for addressing the clean up of the ash spill. In addition, on February 4, 2009, EPA Region 4 and TDEC sent a letter to TVA notifying TVA that, pursuant to Executive Order (EO) 12088, EPA considers the Kingston spill to be an unpermitted discharge of a pollutant under the Clean Water Act. EO 12088 specifies that when EPA finds an Executive agency in violation of a pollution control standard, upon notice from EPA, that agency shall provide to EPA a plan to achieve and maintain compliance with the applicable pollution control standard. In order to meet the requirements of both the TDEC Commissioner's Order and Executive Order 12088, and to ensure the most efficient and expeditious collaboration between the three agencies, the letter directs TVA to provide copies of all plans, reports, work proposals and other submittals to EPA and TDEC simultaneously. EPA and TDEC are coordinating reviews and approvals of the submittals within our respective authorities. EPA's overall objectives for our review and oversight are to ensure that the clean up protects public health, is in full compliance with all applicable Federal law, proceeds in accordance with sound scientific principles, is done as quickly as possible, consistent with prudent management, and restores the ecosystem.

To facilitate coordination of internal activities, on January 21, 2009, EPA Region 4 formed a Kingston Ash Spill Task Force (Task Force). Senior staffers from the Region's air, water, waste, and laboratory programs are represented on the Task Force to ensure complete and adequate coverage of all issues. Draft plans, products and data produced by TVA and TDEC are reviewed by the Task Force and approval by the Region is coordinated through each of these programs. Members of the Task Force and their staff review data for quality control, participate in site visits and reviews, and have kept in close contact with TDEC and TVA during all phases of the recovery to date. Region 4 is also coordinating with EPA Headquarters. This coordination will continue until the site has been restored.

With respect to ash in Emory River, on February 5, 2009, TVA submitted to EPA and TDEC the draft Phase One Dredging Plan. The Phase One dredging plan was revised, and then approved by both TDEC and EPA on March 19, 2009, after final approval of the associated sampling plan and quality assurance plan for the Phase One dredging operations. Phase One dredging began on March 19, 2009, and involves using a hydraulic dredge and a mechanical dredge to remove the ash from the main channel of the Emory River down to a level of 710 feet above mean sea level. Removal of this material is critical to reopening the channel enough to reduce the potential for upstream flooding which could occur with seasonal high water discharges during the spring. TDEC and EPA have required TVA to develop an extensive monitoring and sampling plan to monitor any releases that might occur during the dredging operation and prevent additional harm to human health or the environment. As the dredging is conducted, if any sampling indicates a release of any toxic substances or a turbidity problem, the agencies will order the dredging to stop until additional measures can be put in place.

Phase One dredging is expected to last for at least several months. Phase Two and Phase Three of the dredging will begin after completion of Phase One. Phase Two dredging will address any remaining ash in the Emory River channel down to the original substrate. Phase Three dredging will address ash in the Swan Pond Embayment and its tributaries. The dredging plans for these later phases have not yet been developed by TVA. EPA and TDEC, as well as other local, state and Federal agencies, will be involved as the plans are prepared. EPA and TDEC will also approve the plans before they are implemented.

EPA and TDEC are also reviewing the overall Corrective Action Plan (CAP) which TVA submitted, pursuant to the Commissioner's Order, on February 27, 2009. The CAP, as submitted, was an initial statement of short-term and long-term plans for recovery of the site and final disposal of the ash, and discussed TVA's initial plan for site assessment, environmental monitoring, protection of water supplies and options for ash disposal. Pursuant to the Commissioner's Order and EPA's authority, TVA's CAP, including any updates, will be reviewed and revised to ensure that the clean up provides continued protection of human health and the environment. As part of the review of the CAP, EPA and TDEC met with TVA on March 19, 2009, to begin revisions of the CAP and to discuss next steps for selection of final disposal sites for the ash, to be located off-site. Plan revisions will involve EPA, TDEC and other local, state and Federal agencies and must be approved by both EPA and TDEC.

EPA recognizes that there are ongoing community concerns regarding the impacts from the ash spill and related cleanup activities. To help facilitate communications, EPA, along with TDEC, ATSDR and the Tennessee Department of Health (DOH) participated in a March 5, 2009, public meeting in Harriman, Tennessee, in which TDEC provided sampling data to the community and residents were able to ask questions and express any concerns to agency representatives. TDEC expects to host additional public meetings while the cleanup process continues. We also understand that TVA has planned a public meeting for March 30, 2009, and we encourage TVA to continue efforts to reach out and involve the affected citizens of the surrounding community in the planning and conduct of the clean up.

CONCLUSION

EPA will use its authorities and expertise to continue oversight and technical assistance efforts to protect human health and the environment during the clean up of this incident and promote the restoration of the surrounding ecosystem. EPA will continue to work with other agencies to share information with the community, and will keep Subcommittee staff informed on progress related to the response. Again, we appreciate the opportunity to testify today and will be pleased to answer any questions you may have.

In: Coal Combustion Waste: Management and Beneficial Uses ISBN: 978-1-61728-962-0
Editor: Daniel D. Lowell

Chapter 6

TESTIMONY OF BARRY BREEN, ACTING ASSISTANT ADMINISTRATOR, OFFICE OF SOLID WASTE AND EMERGENCY RESPONSE, U.S. EPA, BEFORE THE SUBCOMMITTEE ON WATER RESOURCES AND THE ENVIRONMENT, HEARING ON "COAL COMBUSTION WASTE STORAGE AND WATER QUALITY"

Madam Chairwoman and members of the Subcommittee, thank you for the opportunity to testify on the U.S. Environmental Protection Agency's (EPA's) coal combustion regulatory development activities and management unit assessment efforts. My testimony provides a brief history of EPA's regulatory efforts on coal combustion residuals, as well as an update on our current rulemaking activities. I will also discuss and provide an update of EPA's assessment of coal combustion residuals management units.

REGULATION OF COAL COMBUSTION RESIDUALS

Coal combustion residuals (CCR) are one of the largest waste streams generated in the United States, with approximately 131 million tons generated in 2007. Of this, approximately 36% was disposed of in landfills, 21% was disposed of in surface impoundments, 38% was beneficially reused, and 5% was used as minefill. In comparison, EPA's Biennial Hazardous Waste Report shows that approximately 33.7 million tons of hazardous waste was generated in the United States in 2007. CCR typically contain a broad range of metals, including arsenic, selenium, and cadmium; however, the leach levels, using EPA's Toxicity Characteristic Leaching Procedure (TCLP), rarely reach the Resource and Conservation Recovery Act (RCRA) hazardous waste characteristic levels. Due to the mobility of metals and the large size of the typical disposal unit, metals (especially arsenic) may leach at levels of potential concern from impoundments and unlined landfills.

The beneficial use of CCR provides environmental benefits in terms of energy savings, greenhouse gas emission reductions, and resource conservation. In 2007, 56 million tons of

CCR were reused. For example, use of CCR contributed to the construction of the Hoover Dam, the San Francisco-Oakland Bay Bridge, and the new I-35 bridge in Minneapolis, Minnesota. Many state environmental statutes and regulatory programs, as well as state road construction agencies, provide for the beneficial use of CCR. In 2007, use of coal fly ash as a substitute for Portland cement in concrete reduced energy use in concrete manufacturing by 73 trillion British thermal units (BTUs), with associated greenhouse gas emission reductions estimated at 12.5 million tons of carbon dioxide equivalent (MTCO2).

Regarding EPA's regulatory efforts for CCR, in May 2000, EPA issued a "Regulatory Determination on Wastes from the Combustion of Fossil Fuels" which conveyed EPA's determination that CCR did not warrant regulation as hazardous waste under Subtitle C of RCRA. However, EPA also concluded that federal regulation as a non-hazardous waste under Subtitle D of RCRA was warranted. EPA based this determination on a number of important findings: (1) the constituents present in CCR include metals that could present a risk to human health and the environment under certain conditions; (2) EPA identified 11 documented cases of proven environmental damage due to improper management of CCR in landfills and surface impoundments; (3) many sites managing CCR lacked controls, such as liners and ground water monitoring; and (4) while state regulatory programs had shown improvement, gaps still existed. With respect to other uses, EPA determined that beneficial uses of CCR, other than minefilling, did not pose a risk and thus did not require federal regulation. EPA also determined that minefilling should be regulated under RCRA Subtitle D or the Surface Mining Control and Reclamation Act (SMCRA).

After the Regulatory Determination, EPA continued to collect new information and conduct additional analyses as part of its effort to develop regulations. In August 2007, EPA made this information available for public comment through a Notice of Data Availability. This notice solicited comment on three documents – an updated draft risk assessment characterizing potential human and ecological risks associated with disposal of CCR in surface impoundments and landfills; an updated report on damage cases associated with disposal of CCRs, which identified an additional 13 proven damage cases; and a Department of Energy / EPA survey of recent disposal practices. In addition, EPA also made available for comment two alternative management approaches, one recommended by a consortium of environmental groups and the other by the utility industry. The comment period on the notice closed on February 11, 2008. EPA received close to 400 comments. After the comment period closed, EPA commissioned a peer review of the draft risk assessment which was completed in September 2008.

The failure of an ash disposal cell at the Tennessee Valley Authority's (TVA's) Kingston plant in December 2008 highlighted the issue of impoundment stability. Our previous regulatory efforts had not included this element; however, we are now analyzing and considering whether to specifically include impoundment integrity as part of our CCR regulatory development.

EPA is committed to issuing proposed regulations for the management of CCR by the electric utilities by December 2009. We are currently evaluating a number of different approaches for regulating CCR, including revising the May 2000 Regulatory Determination. As part of our efforts, we are reviewing all of the information we have on CCR, including all of the comments received from our August 2007 NODA and the peer review of the risk assessment.

REGULATION OF WATER DISCHARGES

Wastewater discharges from surface impoundments are subject to Clean Water Act regulations implemented through the National Pollutant Discharge Elimination System (NPDES). NPDES permits incorporate technology-based effluent limits (i.e., effluent limitations guidelines), water-quality based effluent limits, and standard and special conditions.

NPDES regulatory requirements that address impoundment integrity include standard permit conditions to "...properly operate and maintain all facilities and systems of treatment and control (and related appurtenances)...to achieve compliance with the conditions of this permit" [See 40 CFR part12.41(e)]. In addition, best management practices can be incorporated in NPDES permits as necessary to achieve limitations or to carry out the purpose and intent of the Clean Water Act [See 40 CFR part 122.44(k)].

EPA reviewed a sample of existing NPDES permits to see what types of conditions were currently in permits to address impoundment integrity. EPA determined that additional technical assistance is needed to help permit writers better address coal ash impoundment integrity. As a result, EPA is developing model permit language and implementing guidance that will be discussed with our state counterparts and then made available for state and EPA permit writers. EPA also is considering technical assistance for permit writers to help them identify and apply appropriately sensitive analytical test methods to effectively measure the impacts of both permitted discharges and any future spills.

The effluent limitation guidelines for steam electric power plants were last issued in 1982 and are codified in Part 423 of the *Code of Federal Regulations* (40 CFR part 423). Since 2005, EPA has conducted an intensive review of wastewater discharges from coal-fired power plants to determine whether new Clean Water Act regulations are needed. As part of this effort, EPA sampled wastewater from surface impoundments and advanced wastewater treatment systems, conducted on-site reviews of the operations at more than two dozen power plants, and issued a detailed questionnaire to thirty power plants using authority granted under section 308 of the Clean Water Act. EPA's data collection efforts focused on three target areas: (1) identifying treatment technologies for the wastewater generated by newer air pollution control equipment; (2) characterizing the practices used by the industry to manage or eliminate discharges of fly ash and bottom ash wastewater; and (3) identifying methods for managing power plant wastewater that allow recycling and reuse, rather than discharge to surface waters. EPA has engaged in extensive dialogue with our state partners to ensure their comments about power plant discharges are taken into account.

In August 2008, EPA published an interim report describing the status of the detailed study and findings to date. Much of the information EPA collected, including the laboratory data from sampling and the questionnaire data were made available to the public. The study is still in progress and in December 2008 EPA received the laboratory results from its most recent sampling event. Upon completion of the study this year, EPA will determine whether the current national effluent limitations guidelines for power plants need to be updated. EPA's interim study report, *"Steam Electric Power Generating Point Source Category: 2007/2008 Detailed Study Report,"* can be found online at http://epa.gov/waterscience/guide/ 304m/2006/steam-interim.pdf.

Assessment Efforts

As noted previously in my testimony, the failure of an ash disposal cell at TVA's Kingston plant in December 2008 highlighted the issue of impoundment stability. As a result, EPA has embarked on a major effort to assess the stability of those impoundments and other management units which contain wet-handled CCR. This assessment has three phases: information gathering through an information request letter; site visits or independent assessments of other state or federal regulatory agency inspection reports; and final reports and appropriate follow up. Currently, we are still in the information gathering phase and plan to begin field work in May of this year.

On March 9, 2009, EPA sent information request letters under the authority of the Comprehensive Environmental Response, Compensation, and Liability Act (CERCLA) 104(e) to 162 facilities and to 61 utility headquarters offices. These information requests asked specific questions related to the stability of the management units and required a response within ten working days of receipt. Further, in order to emphasize the priority placed on this effort, EPA's Administrator signed the cover letter for each of these requests. I am happy to report that all of the corporate responses and all but two of the individual facility responses have been submitted. We are following up with those facilities that have not responded. In addition, through this effort an additional 43 facilities with impoundments or management units for wet-handled CCR have been identified. EPA has sent information requests to these facilities. Overall, the assessment responses have identified more than 400 management units that have free liquid.

EPA is in the process of analyzing these responses to determine the appropriate next step for each facility. We plan to conduct assessments for all of these facilities on a case-by-case basis and are evaluating the best methods for conducting these assessments. EPA has retained a contractor to assist in the assessments and we plan to have our first teams in the field in May. We will work closely with our state partners on the scheduling of any site assessments and our state partner agencies will be invited to participate.

If our assessments indicate that corrective measures are needed, EPA will work closely with our state partners to ensure that these measures are taken. In addition, EPA expects to prepare a report for each of the units assessed and make those reports available to the public. Our goal is to complete all of the assessments this year. We will continue to share information about our assessment efforts as they progress.

EPA also is evaluating CCR disposal practices at coal-fired power plants to determine if these facilities are in compliance with existing federal environmental laws and will take enforcement action, where appropriate, to address serious violations.

Conclusion

EPA will continue its regulatory development process and its management unit assessment efforts and we will continue to keep the Committee informed on progress related to these efforts.

In: Coal Combustion Waste: Management and Beneficial Uses ISBN: 978-1-61728-962-0
Editor: Daniel D. Lowell

Chapter 7

JOINT TESTIMONY OF ERIC SHAEFFER, DIRECTOR, ENVIRONMENTAL INTEGRITY PROJECT AND LISA EVANS, ATTORNEY, EARTHJUSTICE, BEFORE THE SUBCOMMITTEE ON WATER RESOURCES AND THE ENVIRONMENT, HEARING ON "COAL COMBUSTION WASTE AND WATER QUALITY"

Thank you, Mr. Chairman, for the opportunity to testify before the Subcommittee on Water Resources and Environment today. My name is Eric Schaeffer, and I am Director of the Environmental Integrity Project, a nonprofit and nonpartisan organization that advocates for more effective enforcement of federal environmental laws. I also served as director of the USEPA's civil enforcement program from 1997 to 2002. The testimony that follows is offered on behalf of myself and my colleague Lisa Evans, a senior attorney at Earthjustice and one of the nation's leading experts on coal ash. Our testimony will make the following points:

1) Coal ash is a hazardous material that tends to leak toxic metals into groundwater and surface water, especially when the ash is saturated or stored in wet ponds.
2) The discharge of wastewater from coal ash ponds, as well as the runoff from so-called dry landfills, can release arsenic, selenium and other pollutants in amounts known to be toxic to human health and aquatic life in our rivers and lakes. Despite the risks, discharges of toxic metals are generally not restricted under Clean Water Act permits at power plants and are often not even monitored.
3) Air pollution control equipment installed to comply with the Clean Air Act will generate thousands of tons of scrubber sludge at a typical power plant. USEPA and industry data show that the wastewater discharged from scrubber sludge treatment systems can release toxic metals like selenium in concentrations that are hundreds of times higher than water quality standards designed to protect aquatic life.
4) USEPA has promised to develop federal safeguards for the disposal of coal ash, but is also evaluating whether to set limits on the toxic discharges from ash and sludge

treatment systems. The monitoring data indicate that such limits are overdue, and there is little time to lose.

COAL ASH IS HAZARDOUS

Coal contains toxic metals like arsenic, boron, cadmium, chromium, lead, and selenium. The National Research Council (NRC) observed in a 2006 report, *Managing Coal Residue in Mines,* burning coal increases the concentration of these pollutants; if the ash is saturated, these pollutants are likely to leak into groundwater or surface water. The NRC examined the growing practice of depositing ash in mines to reduce acid runoff and warned that, "the presence of high concentration levels in many leachates may increase the health or environmental risks near some mine sites." In fact, the USEPA has determined in recent reports that coal ash, when tested with a reliable leach test, exceeds the toxicity characteristic (the threshold for a hazardous waste determination) under the Resource Conservation and Recovery Act for both selenium and thallium.

Most ash is disposed of in landfills or in large ash ponds like the one that collapsed at the Tennessee Valley Authority's Kingston plant in Tennessee just before Christmas. While catastrophic releases remain a real risk at some disposal sites, the leak or discharge of toxic metals from the sites is a daily event at many locations. The USEPA has identified at least 67 proven or likely instances in which groundwater, creeks, wetlands or lakes have been seriously contaminated by arsenic, boron, selenium, and other metals released from ash disposal sites.

Many additional confirmed cases of contamination from coal ash are not on the USEPA's list, including ones that resulted in the destruction of drinking water supplies; the Agency acknowledged in 2000 that the threats from coal ash are likely to be far larger, due to the lack of monitoring at so many coal ash sites. For example nearly two- thirds of the ash ponds in America did not have groundwater monitoring as of 1999, and little has changed since then to require monitoring at these sites.

The U.S. electric power industry generates about 130 million tons of ash, scrubber sludge and other combustion residues annually according to the USEPA , or about 1,000 pounds per person. This volume of waste would fill 1 million train cars, and USEPA predicts that volume will swell to some 175 million tons annually in just six more years. That's comparable to the amount of household garbage that we generate in the U.S. every year, with one important difference: in most states, municipal landfills are subject to significantly more regulation that coal ash dump sites. Leaks from these unregulated operations may not only contaminate drinking water wells, but can also reach rivers and streams through adjacent aquifers.

DISCHARGE OF TOXIC METALS FROM COAL ASH

While toxic metals held in ash ponds, landfills, and treatment systems can leak into groundwater, the wastewater residue from such operations is also routinely discharged into wetlands, creeks, rivers and lakes. Based on annual industry reports to the USEPA's Toxics Release Inventory, power plants are the second largest discharger of metals and metal compounds, releasing more than 2 million pounds in 2008. The actual volume may be significantly larger, since these discharges are not regulated by the USEPA, and are not routinely monitored or reported at many plants.

Our analysis of the limited data that are available through the USEPA indicates that power plants routinely discharge some toxic metals – particularly selenium – in concentrations that exceed water quality standards. For example, selenium is a toxic pollutant found in coal ash that is deadly to fish, and which can also damage the liver and other soft tissues in humans. USEPA has determined that chronic exposure to selenium at levels above 5 micrograms per liter – or about 5 parts per billion -- is harmful to freshwater fish and other aquatic life. Some states have also adopted standards to limit acute (short-term) exposures to no more than 20 micrograms.

Data compiled from permit applications, monitoring reports, and sampling conducted for the USEPA identified at least thirty sites in which routine long-term discharges of selenium exceed 20 micrograms, and sometimes 100 micrograms (See Table A and Attachment A). Selenium water quality standards are meant to protect receiving waters, and do not necessarily apply to the actual discharge of wastewater from pipes. But we have already learned the hard way that releasing selenium into rivers and lakes can decimate fish populations and make the surviving species unsafe to eat. For example, according to the USEPA, the discharge of selenium from a power plant wiped out 16 of 20 fish species in Belews Lake in North Carolina in the 1980s, while selenium contamination from Texas power plants in approximately the same decade led the state to recommend limiting consumption of fish.

Discharge of Toxic Metals from Scrubber Sludge

U.S. power plants that haven't already done so are scrambling to install scrubbers to reduce emissions of sulfur dioxide, in anticipation of Clean Air Act deadlines or to comply with enforcement increases. That is a welcome trend, since scrubbers can remove 95% of the sulfur compounds that cause acid rain and promote formation of fine particles that trigger asthma attacks, heart disease, and premature death. Less welcome is the news that alarming amounts of some of the metals that are stripped out of the smokestack are ending up in our waterways.

Scrubbers generate sludges that need to be periodically treated or dewatered to remove contaminants and reduce the need for additional landfill space. The limited monitoring data available from the USEPA show that selenium levels in wastewater that is discharged from scrubber sludge can be sky-high, reaching concentrations in excess of 1000 parts per billion, or hundreds of times higher than the USEPA's recommended water quality standard of 5 parts per billion.

Release of Arsenic and Other Pollutants

The limited monitoring data available show that power plants also release other pollutants at levels that exceed drinking water standards or limits meant to protect recreational uses like swimming and fishing. The USEPA has established a maximum contaminant level of 10.0 micrograms per liter for arsenic in drinking water. States like Tennessee use the same threshold in waters used for recreational purposes, recognizing that arsenic becomes increasingly concentrated as it moves up the food chain, which could potentially make some fish unsafe to eat. USEPA data identify at least 20 power plants where arsenic levels in wastewater discharges routinely exceed 20 micrograms per liter, or at least twice the recommended federal standard for drinking water or recreational waters. Again, this is likely an understatement, as so few monitoring data actually exist.

EPA Needs to Regulate before it Is Too Late

Air pollution controls create mountains of ash and sludge, and these already staggering volumes will grow rapidly as companies move to comply with new Clean Air Act requirements. But cleaner air should not mean dirtier water, and the USEPA needs to establish strict standards to make sure that we are not just trading one problem for another.

- After decades of delay, the USEPA has promised to propose standards for safe disposal of coal ash no later than the end of this year. Those standards should recognize that coal ash is a hazardous waste. In addition, those standards should apply to scrubber sludges and other types of combustion residue, and address potential risks to both human health and the environment. In particular, the regulations should prevent both the contamination of drinking water, and the pollution of surface waters from adjacent aquifers from both existing and retired coal ash dump sites.
- USEPA standards should also apply to the disposal of coal ash in mines, quarries and other sites that have escaped virtually any common sense safeguards due to exemptions in state laws that are exploited in the absence of federal action.
- Wet storage of coal ash should be phased out as quickly as possible, as the highest threats to human health and the environment occur when coal ash is placed in water.
- USEPA is evaluating the need to set limits on toxic discharges from coal plants – the data it has gathered so far, and the expected growth in waste from new air pollution control equipment, indicate that there is little time to lose. USEPA should move immediately to require more extensive monitoring of the discharge of arsenic, selenium, and other toxic pollutants from power plants and should set discharge limits, including zero discharge limits, consistent with water quality criteria for toxic substances.
- In at least some cases, power plants may be violating federally enforceable permit requirements or rules that limit discharges that contribute to a violation of water quality standards. USEPA's enforcement program, working with state agencies, should investigate and take action where serious violations can be established.

Thank you again for the opportunity to testify and for your attention to this important issue, and I will be pleased to answer any questions that you may have.

In: Coal Combustion Waste: Management and Beneficial Uses ISBN: 978-1-61728-962-0
Editor: Daniel D. Lowell

Chapter 8

TESTIMONY OF DAVID C. GOSS, AMERICAN COAL ASH ASSOCIATION, BEFORE THE SUBCOMMITTEE ON WATER RESOURCES AND THE ENVIRONMENT, HEARING ON "COAL COMBUSTION WASTE STORAGE AND WATER QUALITY"

Madame Chairman, Members of the Committee and Distinguished Panelists:

My name is Dave Goss, former Executive Director of the American Coal Ash Association (ACAA) and I have been asked to appear before you today by ACAA's current Executive Director and its membership. ACAA promotes the recycling of coal combustion products (or CCPs) which include fly ash, bottom ash, boiler slag and air emission control residues, such as synthetic gypsum. It is our opinion, that the U.S. Environmental Protection Agency (EPA) regulatory determinations, made in 1993 and reaffirmed in 2000, are still correct that CCPs DO NOT warrant regulation as hazardous waste.

The recycling of these materials is a tremendous success story that has displaced more than 120 million tons of greenhouse gases since 2000. During that same period, more than 400 million tons of CCPs have been recycled in road construction, architectural applications, agriculture, mine reclamation, mineral fillers in paints and plastics, wallboard panel products, soil remediation and numerous other uses that would have required other materials if these CCP products were not available. Use of 400 million tons of CCPs displaces enough landfill capacity to equal 182 billion days of household trash.

The use of CCPs goes back more than forty years. In the last three decades, the EPA, other federal agencies, numerous universities and private research institutes have extensively studied CCP impact on the environment. The U.S. Department of Energy and the U.S. Department of Agriculture have both funded, conducted and evaluated mining and land case studies using a variety of applications. Consistently, these federal agencies found that when properly characterized, managed and placed, CCPs do not have a harmful impact on the environment or on public health.

EPA reported to Congress on March 31, 2009, results of data collected and analyzed by the Agency from the Tennessee Valley Authority ash spill on December 22, 2008. This data showed that there were no exceedances to drinking water or air quality standards. This information was based on hundreds of water samples and more than 26,000 air samples.

State Departments of Transportation, using technical and environmental guidance issued by the American Society for Testing and Materials (ASTM), the U.S. Federal Highway Administration and the American Association of State Highway and Transportation Officials (AASHTO), have used millions of tons of CCPs without incident or risk. Many years of monitoring and studies following the use of CCPs in road construction have not identified any cases where there has been a negative impact on public health or on the environment.

A goal of this committee, I believe, should be to understand how the use of CCPs has had and can continue to have a positive impact on our nation's resource conservation goals. CCPs have been and should remain a key part of resource conservation efforts because CCPs safely used in lieu of earth, clays, aggregates or soils promote a zero waste goal. Fly ash, bottom ash and synthetic gypsum used to displace the production of portland cement reduce significant carbon dioxide emissions and similiarly conserve natural resource consumption (i.e., the need for quarrying shale, clays or rock gypsum). International and domestic protocols recognize the greenhouse gas reduction benefit of using these materials.

When fly ash is used in concrete, it produces longer lasting, more durable structures and pavements. The fly ash is not just a substitute recycled product; it improves the performance of the concrete. Nearly half of the concrete placed in the U.S. incorporates fly ash because it makes concrete better. We need Congressional support to promote a green supply chain promoting higher replacement rates of fly ash and broader usage. Building longer lasting concrete structures by using fly ash allows our country to move toward a greener and more sustainable economy -- less rebuilding in the future, lower life cycle costs and fewer CO_2 emissions.

A key part of the strategy of recycling industrial materials must be to minimize the need for landfills or disposal facilities. By recycling fly ash in concrete, we bind the fly ash into a concrete matrix and significantly eliminate the potential for any impacts on water resources. Beneficial use regulations are crafted at the state level to promote recycling and to accommodate local environmental conditions. Regulatory programs and policies, developed and implemented by the states, provide for the proper use of CCPs.

The recycling of nearly 43 percent of the 130 million tons of CCPs produced annually is an excellent example of environmental stewardship and sustainability. An effort by EPA or Congress to designate coal ash as hazardous, even if only for the purposes of disposal, could have the dramatic impact of eliminating nearly all these safe, beneficial uses. As America joins the world in seeking to address climate change, a hazardous designation would significantly handicap America as it would not use and therefore not be able to rely upon CO2 reductions from the use of CCPs in lieu of portland cement or other applications. Also, America would have to find environmentally safe disposal facilities for 130 million tons or more of CCPs produced annually. Producers and end-users would no longer use CCPs because of the stigma that a "hazardous" designation would have upon the end user. Furthermore, recycling would end due to the "cradle to grave" liability associated with a "hazardous waste" label.

If this nation is going to develop a culture where safe use and reuse of products and waste streams conserves our nation's resources, CCPs have played and should continue to play an important role in sustainability. Ample technical guidance is available to ensure the environment is protected while still recycling millions of tons of these mineral resources. State specific regulatory guidance will best be able to address local conditions.

As part of the recent economic stimulus efforts supported by the President and Congress, green building has been highlighted. ACAA believes a key component must be the creation of a green supply chain. Developing green jobs as part of a green supply chain and implementing projects that include safe recycling of CCPs should be a vital part of these sustainable projects. With an emerging focus on greenhouse gases, recycling of CCPs contributes measurably to reduction of CO_2 and should, therefore, be encouraged more aggressively. We must better manage our scarce natural resources by using and recycling our existing industrial resources, including CCPs.

Thank you for this opportunity to address this committee.

David Goss

In: Coal Combustion Waste: Management and Beneficial Uses ISBN: 978-1-61728-962-0
Editor: Daniel D. Lowell

Chapter 9

STATEMENT OF THE HONORABLE JIM COSTA, CHAIRMAN, SUBCOMMITTEE ON ENERGY AND MINERAL RESOURCES, OVERSIGHT HEARING: "HOW SHOULD THE FEDERAL GOVERNMENT ADDRESS THE HEALTH AND ENVIRONMENTAL RISKS OF COAL COMBUSTION WASTE?"

Today's hearing is the first time in at least a decade that this Committee has focused on the important issue of coal combustion waste management. I expect, however, that this will be just the beginning of our examination of coal waste. Although our Committee's chairman, Mr. Rahall, has sought solutions to the problem of coal waste management since the 1980s, many of us on this Subcommittee are just beginning to learn about the environmental and health risks of coal combustion waste, and options for its safe management. We intend to hold additional hearings on coal combustion byproducts in which we can gain input from other perspectives, including federal agencies like the Environmental Protection Agency and the Office of Surface Mining, on how best to address the waste challenge safely and sustainably.

Why hold this hearing now? First, because the problem of how to handle coal combustion waste is growing. Coal is a fundamental part of our present and future energy supply. Coal-fired power plants generate half the nation's electricity. But, they yield approximately 125 million tons of coal waste a year that we must reuse or dispose.

Secondly, the time is ripe for this hearing because recent reports raise serious questions about the management of coal byproducts, like fly ash. The Environmental Protection Agency has identified 67 cases in which human or ecosystem health have been compromised by coal combustion waste. And, the Agency's draft risk assessment from 2007 revealed risks to human health and the environment from the disposal of coal waste in landfills and surface impoundments.

Another important report was published in 2006. At Chairman Rahall's request, the National Research Council analyzed how to safely manage coal combustion residues in mines. The Council's report determined that coal waste may cause problems at or near some mine disposal sites, and found gaps and inadequacies in state regulatory programs for coal waste disposal. The report recommended enforceable federal standards for mine placement of coal waste.

In short, today's hearing is an opportunity to gain a better understanding of the dangers coal waste can pose if mismanaged, and get input on what regulation is needed for coal waste disposal--whether in landfills, mines, quarries, or other kinds of sites.

I also think it is important that we examine how we can promote reuse of coal waste in products like concrete and roads. For example, Wisconsin reuses roughly 85% of its coal waste—the highest rate in the country. Caltrans, in my home state of California, is considered a leader among state transportation agencies because it requires the use of fly ash in concrete paving projects. A typical Caltrans project uses at least 25% fly ash as a replacement for Portland cement. What are the opportunities to minimize the coal waste stream nationwide, as Wisconsin and California are striving to do?

My personal belief is that coal will continue to be a critical part of our energy supply--but pollution from coal waste should not be part of America's future. I look forward to learning how we can ensure that common sense safeguards are in place for people, communities, and water supplies.

In: Coal Combustion Waste: Management and Beneficial Uses ISBN: 978-1-61728-962-0
Editor: Daniel D. Lowell

Chapter 10

TESTIMONY OF MARK SQUILLACE, PROFESSOR OF LAW AND DIRECTOR, NATURAL RESOURCES LAW CENTER, UNIVERSITY OF COLORADO SCHOOL OF LAW, BEFORE THE SUBCOMMITTEE ON ENERGY AND MINERAL RESOURCES, HEARING ON "HOW SHOULD THE FEDERAL GOVERNMENT ADDRESS THE HEALTH AND ENVIRONMENTAL RISKS OF COAL COMBUSTION WASTE?"

The Honorable James Costa
Chairman, Subcommittee on Energy and Mineral Resources
House Committee on Natural Resources
Washington, DC 20515

Dear Congressman Costa:

Thank you for the opportunity to appear before the Subcommittee on Energy and Mineral Resources of the House Committee on Natural Resources. The subcommittee has called this hearing to address the question: "How Should the Federal Government Address the Health and Environmental Risks of Coal Combustion Waste?" Implicit in this question is the concern that coal combustion wastes may contain toxic constituents that pose long-term damage to water supplies and the resources that depend on them.

I have spent most of my professional career working on mining issues, with a particular emphasis on coal mining. I was also a member of the National Research Council (NRC) Committee that was called upon recently to study the disposal of coal combustion residues (CCRs) in coal mines as part of the mine reclamation process. That effort was especially relevant to the question posed by the committee.

I have two recommendations that respond to the question posed by the subcommittee. First and foremost, federal policy should treat the *disposal* of coal combustion residues — whether in coal mines, impoundments or landfills — as the option of last resort. Whenever possible, CCRs should be used for secondary beneficial purposes, and such use should be promoted through incentives for secondary use as well as disincentives for disposal. The NRC Committee recommended that secondary use of CCRs be "strongly encouraged." I would go further and argue that disposal of CCRs in coal mines, landfills, and impoundments should not be authorized unless and until the producer demonstrates a substantial and good faith effort to make the CCRs available for secondary use.

In establishing a presumption in favor of secondary use, it will become important to be clear that disposal of CCRs in a coal mine, in an impoundment, or in a landfill does not qualify. While it may be true in some cases that CCRs can neutralize toxic materials at a disposal site, this fact alone should not be used to justify a beneficial secondary use claim. Beneficial, secondary uses must be new uses of the CCRs that allow the user to avoid the use of some other substitute material. Second, where disposal is allowed, federal standards should be established to ensure that the disposal of CCRs does not cause environmental damage.

Before expanding on these recommendations, let me raise an issue about nomenclature. At the outset, federal policy should avoid accepting the characterization of coal combustion residues as "waste" materials. Calling them wastes suggests that they are something for disposal. In fact, most of these wastes have high values for other purposes. I have used the term "residues" which was the term settled on by the National Research Council Committee on which I served. The Office of Surface Mining has used the term "byproducts," and the EPA, simply "products." Whatever term is used, it is important that federal policy recognizes that, for the most part, they are not wastes and that disposal of these materials in mines, impoundments and landfills should be discouraged.

FEDERAL POLICY SHOULD DISCOURAGE DISPOSAL

CCRs come from various sources at coal-fired power plants. The majority — about 57 percent — comes from fly ash, which is the chief residue from burning finely crushed coal, and which is collected in baghouses and from electrostatic precipitators. Flue gas desulfurization (FGD) material is a residue from the wet and dry scrubbers typically used for reducing SO_2 emissions. FGD materials comprise about 24 percent of the CCRs produced at these plants. Bottom ash is a coarser residue that falls out of the boiler and makes up about 16 percent of CCRs. Finally, boiler slag is a molten form of bottom ash that comes from certain types of furnaces. Boiler slag particles have a smooth, granular surface that are uniform in size. About 3 percent of CCRs are in the form of boiler slag.

CCRs are widely recognized as suitable for a range of beneficial uses. For example, fly ash has cementitious properties that can be used in the production of cement and other construction activities, and is also suitable for use in the production of cement, especially in lightweight concrete products. FGD materials are essentially gypsum (calcium sulfates and sulfites), which is the principle material in the manufacture of wallboard. FGD materials are also used in the production of cement.

Much is being done to promote the secondary use of these and other CCRs. The Coal Combustion Products Partnership (C^2P^2) program, which is a cooperative effort that includes the U.S. Environmental Protection Agency (EPA), the American Coal Ash Association, (ACAA), the Utility Solid Waste Activities Group (USWAG), the U.S. Department of Energy

(DOE), the Federal Highway Administration (FHWA), and the U.S. Department of Agriculture (USDA), does a good job of promoting the Secondary use of coal combustion residues in beneficial applications. *See:* http://www.epa.gov/epaoswer/osw/conserve/c2p2/index.htm

The most recent statistics show increasing use of CCRs for beneficial purposes, but much more can still be done. For example, the ACAA estimates that almost 45 percent of the 72.4 million tons of fly ash produced in 2006 (about 32,423,569 tons) was used in 12 of 15 applications that they tracked. This was a 5 percent increase over the previous year. FGD gypsum production in 2006 was about 12.1 million tons, and of that about 79 percent (or 9,561,489 tons) was used, primarily on the production of wallboard and similar products. This is up 2.5 percent over that of 2005. Bottom ash production was about 18.6 million tons of which 45 percent (or about 8,378,494 tons) was used. This was up 4.5 percent from that of 2005. About 2 million tons of boiler slag was produced in 2006 of which 83 percent (or 1,690,999 tons) was used. This was down from the estimated usage of 96.6 percent in 2005. Boiler slag is used primarily in blasting grit and as roofing granules. Because boiler slag comes from older style cyclone furnaces, boiler slag production is expected to decline as these furnaces are retired.

While the economic incentives for secondary use of CCRs are generally strong, there remains a great deal of CCR disposal that would not likely occur if the true cost of disposal were factored into such decisions. Among the external costs that are unaccounted for in CCR disposal are the societal and economic costs of mining virgin materials, including the carbon footprint from such activities, and the environmental costs and associated risks that result from CCR disposal. While a complete accounting of these costs should be made, these external costs are sufficiently obvious to warrant the immediate imposition of incentives for secondary use and disincentives for disposal of CCRs. This might, for example, include a modest tax on CCR disposal, the proceeds from which could be used to promote secondary use of CCRs. A $0.10/ton tax on the nearly 53 million tons of CCRs that were disposed of in 2006 would yield revenues of $5.3 million, and this money could be used to help establish markets for CCRs or to otherwise incent CCR producers to make secondary use of these materials.

In addition, and as suggested previously, federal and state policies and laws should encourage beneficial secondary use of CCRs by demanding that CCR producers demonstrate a substantial and good faith effort to make the CCRs available for secondary use. This should include an analysis of the suitability of the particular CCRs that are being produced for secondary uses, the relevant markets that might exist for those CCRs, and the efforts that have been made to market those CCRs to interested parties. Federal and state policy could promote these markets by establishing minimum CCR content (or CCR preference standards) for road building materials in Federal Aid Highway projects.

Even as secondary use is encouraged, some CCR disposal will certainly continue, especially in the short term. Because CCRs may contain toxic constituents, the NRC Committee concluded that enforceable federal standards should be established when CCRs are disposed of in coal mines. Logically, the need for such standards applies to CCR disposal in impoundments and landfills as well. The establishment and implementation of these standards is important not only to protect the environment and public health, but also because strict standards will themselves promote the beneficial secondary use of CCRs. Notably, in Wisconsin, which has one of the best programs in the country for managing CCR disposal, 85 percent of CCRs were beneficially used in 2004 as compared with only 35 percent nationally. Coal Combustion Waste Management at Landfill sand Surface Impoundments, 1194- 2004, DOE/PI-0004 (April, 2006)

Among the issues to be resolved regarding federal CCR disposal standards are the questions of which federal agencies should be primarily responsible for managing CCRs, and what standards should be imposed. Once again, the NRC Committee lays out a useful roadmap for answering these questions. The EPA is the federal agency most closely associated with managing waste disposal so it makes sense that the EPA will be significantly involved in this process. Nonetheless, the NRC Committee was focused on CCR disposal at coal mines during the reclamation process, and coal mining reclamation is under the jurisdiction of the federal Office of Surface Mining. Given these overlapping roles, the NRC Committee wisely recognized that coordination between the Office of Surface Mining and the EPA was needed. The Office of Surface Mining will not be involved in CCR disposal in impoundments and landfills, but it makes good sense that mine disposal standards would be consistent with standards for impoundments and landfills. Thus, it is critically important that the EPA be closely involved with the Office of Surface Mining in developing standards for CCR disposal in mines, and that EPA use those standards as a template for federal standards for impoundments and landfills, if Congress grants EPA the authority to promulgate such standards.

As for regulatory standards, the NRC Committee lays out a sensible outline for such standards. Drawing on the Committee's recommendation, Congress should pass appropriate legislation to enforce that the following standards should be implemented at all landfills, impoundments, and mines that are subject to CCR disposal:

1. **CCR and Site Characterization.** Both the disposal site and the CCR materials must be assessed and characterized to determine their potential for promoting leaching of toxic materials on their own and once they are combined at the site.
2. **Site-Specific Management Plans and Performance Standards.** A specific plan must be developed for the disposal at the particular site, and site-specific standards must be established that assure the protection of the environment and public health. Generally, sites should be designed to minimize the flow of water through CCRs so as to minimize the potential for leaching toxic materials.
3. **Monitoring and Bonding.** Given the uncertainties and risks associated with CCR disposal, the placement of a suitable number of monitoring wells should be required with special attention to wells that are down-gradient from the CCR disposal area. An adequate bond or other financial assurance should also be required to assure that the regulatory agency can cover the costs of remedial action, should such action become necessary.
4. **Public Participation.** The public has a strong interest in assuring the disposal of CCRs does not adversely affect the environment or public health. Thus, any CCR disposal proposal should be explicitly made subject to an environmental assessment process with the opportunity for robust engagement of the public on issues of concern.

While much of what I have recommended to the committee can be accomplished without legislation, legislative direction could be very helpful in clarifying federal policy and especially in promoting the beneficial secondary use of CCRs. For this reason, I look forward to an ongoing dialogue with the Committee and its staff as it considers whether legislative action may be necessary or appropriate.

Thank you for opportunity to present these views to the Committee. I welcome your comments and questions.

Sincerely,

Mark Squillace
Professor of Law and
Director, Natural Resources Law Center

In: Coal Combustion Waste: Management and Beneficial Uses ISBN: 978-1-61728-962-0
Editor: Daniel D. Lowell

Chapter 11

TESTIMONY OF SHARI T. WILSON, SECRETARY OF THE MARYLAND DEPARTMENT OF THE ENVIRONMENT, BEFORE THE SUBCOMMITTEE ON ENERGY AND MINERAL RESOURCES, HEARING ON "HOW SHOULD THE FEDERAL GOVERNMENT ADDRESS THE HEALTH AND ENVIRONMENTAL RISKS OF COAL COMBUSTION WASTE?"

Chairman Costa, and honorable members of the Committee, thank you for the opportunity to share Maryland's experience with coal combustion waste with you and, more importantly, for your interest in this very important issue.

We also greatly appreciate Congressman Sarbanes' interest and attention to issues surrounding the disposal of this by-product of producing energy from coal.

In 2006, the most recent year for which complete information is available from Maryland's Public Service Commission, coal generated 60.1% of the electricity generated in the State. In Maryland, there are five companies who generate coal combustion by-products at 9 facilities. Approximately 2 million tons of coal ash (fly and bottom ash) is generated annually from Maryland plants. Of that 2 million tons, approximately 1.6 million tons of coal ash is from the plants owned and operated by two companies, Constellation and Mirant.

In Maryland, the Maryland Healthy Air Act requires flue gas desulphurization equipment (known as "scrubbers") to be put in place by 2010 to reduce sulphur dioxide (SO2) emissions by 80%. A second phase of requirements in 2013 will increase the emission reductions to 85%. That equipment, while reducing SO2 emissions by over 200,000 tons will also increase the volume of scrubber sludge produced by 2.5 million tons. By 2013, therefore, facilities in Maryland will generate 4.5 million tons of CCWs.

As you are aware, coal combustion by-products are frequently reused. Currently, approximately 1 million tons, or one half of the coal ash produced annually, is beneficially used in Maryland. Fly ash can be reused for concrete manufacturing and in building material. It can also be used as structural fill in roadway embankments and development projects. (It can also be used in agricultural applications. While these are just a few of the reuse applications, there are many outstanding questions with regard to the safety of reuse.) For example, when used for structural fill, should liners be used; should there be defined distances between use of CCWs and potable water sources; should it be prohibited in shoreline areas such as the Chesapeake Bay Critical Area, source water protection areas, wetlands, or other areas of special concern; if used in agriculture, should it be applied to crops that are for human consumption. These are issues being examined as the State begins to develop a second phase of regulations to more effectively control reuse.

While reuse is the goal and preferred alternative, currently in Maryland, approximately half of the coal combustion by-products generated in Maryland are disposed of or used in mine reclamation. Maryland has 29 locations where these materials are disposed of or used in mine reclamation.

Currently, in Maryland, regulatory controls exit through mining and/or water discharge permitting authority, but the State currently does not have regulations that are specific to the management and control of CCWs.

At two of disposal sites, within the past year, the Department of Environment has taken legal action to require cleanup of groundwater or surface water contamination. This contamination results from the placement of 4 million tons at one site and 5.5 million cu/yrds at a second site. The groundwater contamination at one site affected residential drinking water wells. As a result, the Department required groundwater remediation, provision of a temporary water supply and eventually a connection for residences to a public water supply. The severity of the situation resulted in the third largest civil environmental penalty in state history, a fine of $1 million.

Prior to that action, the Department began to assess how it regulated the disposal of this material. We were concerned that the regulatory controls Maryland was using needed to be improved given the range of disposal sites and the varying geology and subsurface conditions in Maryland.

At that time on 2007, we were aware that the Environmental Protection Agency (EPA) had been working on regulations since 2000 to institute additional controls on the management of CCWs but had not finalized a proposal. The lack of any federal standard combined with the immediate need to better control disposal prompted Maryland to develop new regulations to strengthen controls on the management and disposal of CCWs. In a very short timeframe, within 8 months, Maryland proposed regulations for public review and comment at the end of 2007 and announced our intent to develop a second set of regulations dealing with the beneficial reuse of CCWs this year. At least two local governments in Maryland have also begun considering the extent to which they should institute, through their land use planning and zoning authority, additional controls.

Developing and implementing regulations such as these also present a new expense for the State. To address that issue, during the legislative session of the Maryland General Assembly, the Department proposed legislation to establish a fee to be paid by a generator of coal CCWs based on a per ton rate of CCWs generated annually excluding CCW that was beneficially reused. While the legislation was not enacted, there was general recognition of the need for the regulations and the need to pay for implementation. The Maryland Department of Environment continues to aggressively work on this important issue using the State resources available to us.

While, we do not believe it is necessary or appropriate to regulate this material as a hazardous waste, clearly, there is a need for more stringent management and control of CCWs in order to protect human health and the environment in Maryland.

We believe there is also a need for action at the federal level. First, a basic premise of the RCRA statute is to promote reuse. There are many opportunities for the federal government, through research, to more effectively assess reuse opportunities and, as a result, to significantly reduce the volume of material that must be disposed. Alternatives to disposal must be maximized to the greatest extent possible.

Second, we believe that the federal government should establish a minimum set of standards for land disposal such as requiring landfill type liners at non-mining reclamation sites as Maryland proposes to do. We are aware that other States, not just Maryland, are dealing with ground and surface water contamination issues from disposal. This is also an area where a threshold of consistency from state to state would be beneficial.

It is, however, critical to note, that with this issue a one size fits all approach will not work. It will not work due to the many variables that control safe disposal such as geology and groundwater characteristics. Each state must be able to tailor standards based on the type of ash generated, the characteristics of that ash, the land disposal methods used, the geology and groundwater conditions and many other characteristics that affect whether disposal is protective of public health.

Thank you for taking the initiative to inquire into this important issue and for the opportunity to share Maryland's perspective.

In: Coal Combustion Waste: Management and Beneficial Uses ISBN: 978-1-61728-962-0
Editor: Daniel D. Lowell

Chapter 12

STATEMENT OF DAVID GOSS, EXECUTIVE DIRECTOR, AMERICAN COAL ASH ASSOCIATION, SUBCOMMITTEE ON ENERGY AND MINERAL RESOURCES, HEARING ON "HOW SHOULD THE FEDERAL GOVERNMENT ADDRESS THE HEALTH AND ENVIRONMENTAL RISKS OF COAL COMBUSTION WASTE?"

Good morning, Mr. Chairman. My name is David Goss, Executive Director of the American Coal Ash Association. I sincerely appreciate the opportunity to address you, the members of the Committee and other distinguished experts appearing before you on this important topic. ACAA is an industry association of producers, marketers, end-users, researchers and others who support the beneficial use of what our industry refers to as coal combustion products, commonly known as CCPs. This includes coal ash and residues from air emission control systems such as synthetic gypsum products. These materials are the residuals from the burning of coal to generate electricity. By the very nature of the energy generation process utilizing coal, these byproducts cannot be eliminated entirely and must be managed like many other industrial byproduct streams. We consider CCPs to be mineral resources that if not used, become resources that are wasted.

In a perfect world, energy generation would not have any byproducts because the process would efficiently use all of the raw materials needed to generate electricity. Yet, the coal fueled generation process is not perfect. Even other energy options have consequential impacts, for example wind, which yields noise pollution and bird impingement. The coal-based energy generation industry generates byproducts including fly ash, bottom ash, slag and gypsum. The difference is that many of our products can replace or improve other commonly used commodities including portland cement and constituents which are used to produce concrete and other construction materials. The safe re-use of CCPs has a significant positive impact on this nation's mineral resources, its environment and economy. It is essential to promote and support activities that contribute to a more sustainable nation. By sustainable nation, I mean efficient, socially responsible and environmentally friendly usage of CCPs. I think the majority of us would agree that byproduct re-use which is environmental, health and safety conscience is much better than putting wastes in a disposal facility. Recognizing this common interest to promote safe and environmentally sound byproducts use, I am here to address how the beneficial use of CCPs contributes measurably to reduce environmental impact and is properly being regulated by the federal and state authorities.

BACKGROUND INFORMATION

Annually, more than 125 million tons of CCPs are produced and more than 54 million tons (or 43%) are used beneficially. These beneficial uses include: raw feedstock for portland cement production... as a replacement for portland cement in concrete and concrete products... as mineral filler in asphalt... as aggregates in road construction... for soil modification and stabilization... .for wallboard panel products... in agriculture...in coal mine reclamation and many other commonly accepted uses.

The premise of this hearing is what should be done by the federal government to regulate CCPs. I believe that the federal government has for years worked closely with states to address the impact of CCPs in all media: water, land and air. I am taking the liberty of highlighting only a few of more recent federal efforts. Our industry believes this partnership between federal and state authorities has allowed state governments to remain agile to address unique issues related to local topography, climatology and land conditions (including abandoned mine lands). We do not see a need for this regulatory balance to be legislatively adjusted at this time.

On May 22, 2000, the United States Environmental Protection Agency ("EPA") confirmed in the Federal Register that regulation of CCPs under Subtitle C of the Resource Conservation and Recovery Act, ("RCRA") was not warranted. Furthermore, the EPA stated "we do not want to place any unnecessary barriers on the beneficial uses of these wastes, because they conserve natural resources, reduce disposal costs and reduce the total amount of waste destined for disposal." The EPA also stated, "We have not identified any other beneficial uses that are likely to present significant risks to human health or the environment and no documented cases of damage to human health or the environment have been identified." (See 65 Fed. Reg. 32214 to 32228, May 22, 2000).

In 2004, the United States Department of Energy ("DOE") and EPA issued a detailed evaluation of the placement of CCPs in landfills and surface impoundments, for the period 1994 through 2004. This study was done to provide additional information not available during the regulatory determination process that supported the position taken by the EPA on May 22, 2000 cited above. The report concluded that the information reviewed showed improved management of CCPs was seen in both landfills and surface impoundments. Additionally, 100% of the sites reviewed were covered by one or more state issued permits.

During 2004 and 2005, the National Research Council of the National Academies conducted an extensive evaluation of the use of CCPs in mining activities, the results of which were published in 2006. The committee concluded that the use of CCPs as part of mine reclamation is appropriate provided that an integrated process of characterization, management and engineering design is in place to reduce potential risks. Because of this conclusion and the other recommendations by the committee, the Office of Surface Mining

("OSM"), in consultation with the EPA, is taking the lead role in developing proposed rulemaking. The OSM rules would pertain to permit applications and performance standards for coal mine reclamation under Title V of the Surface Mining Control and Reclamation Act of 1977 ("SMCRA" or the "Act") or in the reclamation of abandoned coal mine sites funded under Title IV of the Act. This rulemaking is anticipated to be issued in the summer of 2008.

BENEFICIAL USE

Mr. Chairman, it is our opinion that the current state and federal regulatory process is more than adequate to protect both the environment and to address any potential health risks to the general public. Recently there was a situation in Anne Arundel County, Maryland where the placement of CCPs (at the Gambrill's site) was found to be impacting local groundwater. As a result of that incident, the State of Maryland immediately intervened,, operations were halted and worked with the company involved and the local community to correct the situation. Furthermore, the Maryland Department of the Environment ("MDE") has instituted a full review of their solid waste and beneficial use regulations as they pertain to CCPs. The lessons learned at this one site are being shared with surrounding states and with other states through the Association of State and Territorial Solid Waste Management Officials (ASTSWMO) and EPA regional offices to understand the specific situation at this location. This unusual situation, in our opinion, does not warrant broad federal regulations.

The Commonwealth of Virginia has just formed a Technical Review Committee to assess the adequacy of the State's current CCP regulations along with a broader review of Virginia solid waste regulations. The first meeting of this broad based advisory group is scheduled for later this week. This regulatory review process will identify any situations or scenarios where changes to Virginia regulations might be needed.

It is our opinion that most states want to continue their role in the oversight of management, recycling and beneficial use of CCPs and other industrial byproduct streams. Routinely conducted for many years, industrial recycling of materials continues to play an important role.

Gambrill's is, we think, an isolated example related to one CCP situation. As discussed above, other surrounding states are looking at these circumstances to ensure any lessons learned are instituted to protect their citizens and environment. In 2006, more than 54 million tons of CCPs were used in fifteen application categories. These include use in concrete and concrete products; the production of portland cement; flowable fill materials; structural fills and embankments; road base and soil modifications; mining, agricultural and other construction activities. These applications have enabled contractors, end-users and project owners to reduce the consumption of raw materials, helped reduce greenhouse gas emissions and have eliminated the need for new landfill or impoundment space.

Our Association believes that using CCPs in these numerous proven applications is not "disposal." CCP re-use alternatives have been demonstrated by analysis, research, testing and successful construction and remediation activities. For example, it is a measureable benefit that using fly ash in concrete as a partial replacement for portland cement can decrease CO2 emissions and improves performance, strength and durability of the concrete. CCPs do not just replace the portland cement, they improve the product. Increasing the longevity of structures by using fly ash, for example, reduces the need for replacement or re-construction of this nation's transportation and building infrastructures, This exemplifies how beneficial use today can better provide for future generations.

CCPs are also used extensively in coal mine reclamation to help achieve approximate original contour requirements, to eliminate dangerous high walls, as a soil amendment, to neutralize harmful acid mine drainage and for many other beneficial uses. The EPA has evaluated CCPs extensively in the last three decades and continues to affirm they are not hazardous to the public or to the environment when properly managed and used. In the May 22, 2000 regulatory determination, the EPA stated, "There have been no proven damage cases related to postSMCRA placement of CCPs in coal mines."

For use in mining, the OSM, ASTM, DOE and a number of universities have provided technical guidance and have supported research and demonstration projects that have proven that when properly managed and placed, the beneficial use of CCPs can significantly improve conditions at active and abandoned mining sites. The DOE funded Combustion By-Products Recycling Consortium ("CBRC") has issued a number of project reports concerning the use of CCPs in mining and other applications that demonstrate their safe and effective use.

The State of Pennsylvania has documented many cases where the use of CCPs has significantly improved abandoned mine sites within the Commonwealth. Pennsylvania's positive experience with CCPs is fully described in its 2004 publication "Coal Ash Beneficial use in Mine Reclamation and Mine Drainage Remediation in Pennsylvania.

There are a significant number of industry-developed comprehensive technical standards for CCP use that address engineering properties, testing procedures and design considerations (including geological, hydrological and construction techniques). Included in this design process is specific guidance about minimizing environmental impacts such as fugitive dust, groundwater impact and storm water runoff. These documents and specifications detail protections to the environment and the public, as well as specifying quality, technical performance and other criteria. For example, the American Society for Testing and Materials International ("ASTM") has developed several standard and guideline documents that provide technical information on the use of CCPs in structural fills, embankments and mining activities. Additionally, there are many other similar technical documents issued by ASTM, American Concrete Institute ("ACI") and the American Association of State Highway and Transportation Officials ("AASHTO") that address the use of CCPs in road construction, architectural uses, as aggregates, in soil applications and in concrete products.

Furthermore, the Federal Highway Administration ("FHWA"), the DOE, the EPA, the Electric Power Research Institute ("EPRI"), the Recycled Materials Resource Center ("RMRC"), the Turner-Fairbank Technical Center and AASHTO have supported research, conducted studies, provided training and issued technical guidance covering the use of these same CCPs in highway construction, road work and land applications. For example, years of monitoring of highway and road construction projects across the nation have seen no health or safety issues resulting from the use of CCPs. In a study by the RMRC at the University of New Hampshire, it was concluded that:

> Studies and research conducted or supported by EPRI, government agencies, and universities indicate that the beneficial uses of coal combustion products in highway construction have not been shown to present significant risks to human health or the environment.

The practice of using sound management techniques and evaluating the specific project conditions is implemented widely. EPRI, ASTM, ACI, FHWA and state agencies have guidance documents that provide technical and environmental considerations to engineers, contractors and highway authorities on the use of CCPs in highway and road construction and land reclamation. Federal and state agencies routinely approve CCPs for use in road construction because there are well established technical practices that address potential CCP impact on the environment. Some states further define the use of CCPs under their own codes and regulations, further substantiating the beneficial value that CCPs can offer. Other states may not approve all CCPs for use for road construction but welcome the use of fly ash, for example, as a partial replacement for portland cement. These geographic distinctions are worthy of note because they mirror the natural and economic climates and differences that face different states or regions.

In 2003, the EPA, DOE, FHWA and the CCP industry formed the Coal Combustion Products Partnership ("C^2P^2"). This is a nation-wide effort under the Resource Conservation Challenge to help promote the beneficial use of coal combustion products and the environmental benefits that result from their use. The partnership has established a goal of 50% utilization of CCPs by the year 2011, a goal that was mutually agreed upon by the EPA, industry, DOE, the Utility Solid Waste Activities Group ("US WAG") and FHWA. The partnership is fully described at the EPA website http://www.epa.gov/epaoswer/osw/conserve/c2p2/index.htm. This website provides technical and environmental information about using CCPs in ways that conserve natural resources, reduce the need for landfills or disposal facilities and that can reduce greenhouse gas emissions. Case studies and documents describing CCP applications are available to interested parties. C^2P^2 partners include producers, marketers, state agencies, end-users and researchers whose experiences with CCPs further demonstrate the value that these materials can offer.

CONCLUSIONS

We need to use fewer natural resources and use more industrial byproducts to improve our society and sustainable economy. As President Carter stated, "*We simply must balance our demand for energy with our rapidly shrinking resources.*" Naturally, the use of any byproduct must be done in a socially responsible manner that addresses environmental, health and safety needs. We believe that the current federal and state regulatory schemes are well suited to address CCPs use and management.

Regulations affecting air, water and solid waste all have an impact upon industrial practices and resulting byproducts. Air quality requirements are primarily driven or controlled at a federal level. Water and solid waste regulations have been developed at the national and state level since many studies have recognized that risks are not the same all across the country and impacts are better governed at a local level to address specific geological, hydrological or climate conditions.

As described above, key federal agencies including the EPA, the DOE, the FHWA, the OSM, the US Department of Agriculture and along with many states have funded, supported and promoted many beneficial uses for CCPs. Extensively documented research and field projects reinforce our position and theirs that using CCPs is both technically and environmentally sound and provides greater benefit to the environment than disposal.

ACAA and the CCP industry believe that current federal and state regulations are protective of the environment and public health. Most states have developed regulatory guidelines for management and beneficial use for CCPs, which have implemented practical and technically sound methods for managing these materials. When a negative example is found, states intervene and share their experiences through ASTSWMO, EPA regional offices and technology transfer activities that support each state's unique needs. Additional legislative or broad brush regulatory schemes aren't warranted to address an isolated instance.

Years of actual field experience have shown that the benefits of using CCPs in lieu of other materials have not had a negative impact on the environment, public health or safety. Engineering and environmental professionals within private sector, federal and state agencies acknowledge and support the many values of using CCPs.

As your website so clearly states, this nation needs to maintain a healthy balance between providing for energy needs and conserving our nation's precious natural resources. One way which has proven effective is to safely use industrial byproducts such as CCPs. Existing programs and regulations may need to be occasionally adjusted at a federal or state level but wholesale prohibitions on certain re-use applications or new federal regulatory schemes are unwarranted. Existing technical and environmental controls are already available to state and federal agencies to ensure that CCPs will continue to be properly used. The use of CCPs (in conjunction with good engineering judgment and the need to conserve natural resources) can provide many benefits to the public without environmental risk while promoting sustainable construction and infrastructures.

Thank you again for this opportunity to address this committee.

David Goss

In: Coal Combustion Waste: Management and Beneficial Uses ISBN: 978-1-61728-962-0
Editor: Daniel D. Lowell

Chapter 13

WRITTEN TESTIMONY OF CHARLES H. NORRIS, BEFORE THE SUBCOMMITTEE ON ENERGY AND MINERAL RESOURCES, HEARING ON "HOW SHOULD THE FEDERAL GOVERNMENT ADDRESS THE HEALTH AND ENVIRONMENTAL RISKS OF COAL COMBUSTION WASTE?"

I would like to thank Representative Costa and the members of the subcommittee for the opportunity to testify today.

INTRODUCTION

The question the subcommittee is exploring carries important, implicit understandings in its phrasing. There is implicit understanding that coal combustion waste (CCW) exists. There is implicit understanding that there are health and environmental risks with CCW. There is implicit understanding that the risks need be addressed. There is implicit understanding that federal action is needed to address the risks. I share the each of those understandings with the author(s) of the question, although I must admit resistance in reaching the last understanding.

My understandings are founded in 5½ decades of personal observation, management, and study of CCW. In the 1 950s I became responsible for removing, carrying, and dumping the "clinkers" from our coal furnace. They were put to "beneficial use," providing traction and filling ruts on the lane coming up the hill to the farmhouse. In the 1 960s, I became painfully aware that even beneficial use of these materials carries risks, as did everyone else who tried to skate on an icy road after the township trucks had spread cinders or who tripped on the cinder track during the hand-off in the mile relay. In the 1960s and 1970s, I was episodically subjected to the rain of fly ash and the taste and feel of sulfur dioxide in my throat when the wind was from the university's power plant in Champaign, Illinois. Since the mid-1980s, a significant portion of my professional career has been the study and evaluation of CCW, now remove from the air, and how best to manage it. My client base through the years has included individuals, coal companies, environmental organizations, power companies, governmental units, and citizens' groups.

My testimony today represents my personal understanding and opinions, and is not intended to represent those of any other individuals or organizations. My opinions and understanding have evolved and should continue to evolve as I learn more. If they don't, I should retire. I am not being paid to be here and my preparation for this hearing is similarly donated, although I am seeking reimbursement of direct travel expenses.

I will organize my testimony today around the implicit understandings in the question before the subcommittee, largely providing technical background on CCW based upon my personal experience. Consistent with the question before the subcommittee, I will use the term "coal combustion waste." Some of my testimony will touch on language; the nomenclature and rhetorical battle over these materials. That battle contributes to the need for federal intervention to reverse the deplorable and deteriorating conditions manifest under some state management practices for CCW and begins to spread to other waste streams. I will illustrate my points with examples from my own experience and have included studies and research with my testimony to that end. These tend to be lengthy, and some are technically detailed. They are not provided with the expectation that you will fully absorb them. Rather, I hope they will convey the complexity of these materials and of their relationships to and reactions with the environments where they are increasingly placed. Generalization about these materials are difficult, and I hope the supplemental materials help illustrate that.

The difficulty with generalization is seen in the implicit understanding that CCW exists. Certainly the burning coal leaves behind material after combustion; tens of millions of tons of each year. But it's not a single material. There is the first-order classification of these materials as fly ash, bottom ash, flue gas desulfurization (FGD) materials, and boiler slag, each of which is very different. The character and composition of these individual materials are themselves variable. They varies over the range of combustion and pollution abatement technologies that are used. They change as the compositions of the fuels change. They are dependent upon other waste streams that are mixed and co-managed with them. Often, state regulations are broadened to include not just the materials that remain after "coal combustion," but the materials that remain after "fossil fuel combustion." Fossil fuel combustion typically represents a mix of little 50% coal with some other fuel; natural gas, petroleum liquids, wood, wood pulp, shredded tires, auto fluff, etc. There is nothing similar between the FGD sludge produced by a dual-alkaline system working on the stack gases of a pulverized-coal conventional plant burning Wind River Basin coal from Wyoming and the bottom ash from a fluidized bed combustion unit burning 50% coal, 30% gob, and 20% shredded tires. Yet, these two materials, among a host of comparably dissimilar materials, are within the term "coal combustion waste" in the question before the subcommittee, and all need fall under the rubric of federal control.

The challenge at the federal level of addressing health and the environment from risks of this complex of materials does not lie with legislating the detailed management of each material. It lies with producing a framework that provides regulation of each of these materials in a manner that is protective of health and the environment when implemented by state programs. The implementation would be based upon individual CCW characteristics, the nature of its placement and use, the environment of its placement and use, and the time-dependent changes that CCW and the environment work upon each other. The model for this framework is not unlike that of SMCRA or the CWA. I believe producing the framework will be a challenge because such a framework would be a sea change from the approach taken today by some states in response to the systematic hesitation and reluctance of federal regulators to meaningfully regulate these materials.

In the remainder of my comments, I will briefly outline examples of the need for federal intervention based upon what has and is happening under state regulatory programs. I will then provide an outline of issues the federal framework will need to address to be effective. Finally, I will discuss the issues of nomenclature and rhetoric that are driving not only the management of CCW but increasingly undermining the responsible regulation of other wastes streams.

THE NEED

Placement of CCW in the environment creates environmental damage and human health risks. Not every CCW. Not all placements. Not always without some offsetting benefit. However, documented degradation coincident with the placement of CCW in the environment occurs so frequently, in such a wide range of settings, that there must be the presumption that unacceptable risk to health and the environment will occur as a result of such placement. The frequency of such degradation is particularly disturbing when one considers, first, how rarely such placement is accompanied by monitoring at times and places capable of detecting a problem, and second, how frequently the some state agencies ignore the degradation and allow it to continue. Too often, there is no agency response to a problem at all or until affected citizens have had to file legal action for relief. Further, under existing and evolving state programs, the characterization of placement sites is being reduced and monitoring of placed CCW is occurring less often and for shorter periods of time. Intervention to prevent risk to health and the environment from degradation is increasingly impossible because there is no observation.

Examples of degradation are readily found despite the paucity of sites with monitoring data that allow evaluation. The following are some representative examples of the variety and range of problems with in-environment placement of CCW:

Fly ash was placed in an open, unlined excavation as permitted landfill disposal adjacent to Town of Pines in Indiana beginning in the 1970s. Leachate from that ash, passing under the residents' houses, ruined their water supply on its way to local drainage to Lake Michigan, forcing them to accept municipal water as a replacement. The site is undergoing an RI/FS under the SuperFund program.

In Maryland, operators of ash disposal pits are today wringing their hands over ruined residential wells, questioning how they could have ever anticipated such problems from benign materials compliantly disposed.

In Illinois, CCW placement as permitted landfill disposal in a dolomite quarry degrades ground water as a result of off-site, third-party changes to the hydrogeology that had been relied upon to contain the waste.

In Pennsylvania, regulators document ground- and surface water contamination at permitted CCW disposal facilities that in cases rely only CCW for containment. Dilution by the receiving body of water is accepted as a response by the agency.

In Colorado, the USEPA fell victim to beneficial use. Uranium tailings at a site within Denver were "stabilized" using CCW with a liming additive. The objective was to allow reburial of the stabilized tailing on-site, rather than expensive transport and disposal at a rad-waste facility. The "beneficial use" effect lasted only a few months before uranium mobility from the site increased beyond the pre-treatment levels, necessitating the transport and landfill disposal of not only the uranium wastes but also the admixed coal combustion materials.

Contamination examples are also common when CCW is placed in coal mines. The Clean Air Task Force (CATF) contracted an exhausting, multi-year study of the contamination at coal mines that placed CCW as part of the Pennsylvania mining program for beneficial use of CCW. That study found in agency permit files data showing CCW contributions to rising contamination at the majority of the sites with sufficient data to make a determination.

Two of the accompanying documents I am providing with my testimony discuss contamination resulting from the placement of CCW in mines. Some of these placements were beneficial-use placement and some simply disposal placement. In my 2003 report "Minefill Practices for Power Plant Wastes, An initial Review and Assessment of the Pennsylvania System," I discuss my preliminary review of 10 mine sites in Pennsylvania that saw to CCW placement and showed subsequent related contamination. Many of these sites were studied in more detail as part of the CATF study mentioned above. The second paper, "Environmental Concerns and Impacts of Power Plant Waste Placement in Mines," was presented in 2004 and published in Proceedings of State Regulation of Coal Combustion By-Product Placement at Mine Sites: A Technical Interactive Forum, Kimery C Vories and Anna Harrington, editors, by U. S. Department of Interior, Office of Surface Mining. This paper discusses eight mine sites in Pennsylvania (some duplicated in the CATF study), West Virginia, Indiana, and New Mexico where CCW placement can be tied to subsequent ground water contamination.

Where data exist than can be assessed, the frequency of contamination from the placement of CCW is attributable largely to weakness in state programs for site characterization waste and waste characterization. The dearth of interpretable data from most sites is attributable to poor site characterization, poor waste characterization and inadequate monitoring. In the discussion of each below, it should be apparent that the three weaknesses are intimately related.

Monitoring

The first requirement to detect the impacts from the placement of CCW in the environment is a monitoring system and program. One cannot document impacts, or lack of impacts, due to CCW placement without a monitoring program that looks for such impacts and a monitoring system that is capable finding such impacts when they occur. Yet, as more and more CCW is placed under programs of beneficial use, there is an ever-expanding population of placement sites with no monitoring.

To detect impacts from the environmental placement of CCW, a monitoring system must monitor the path(es) of contaminant migrating from the placement area. This requires there have been a site characterization that establishes the migration direction(s), including seasonal variations, of contaminants from the placement area via air, surface water, and ground water. Further, since placement of the CCW can modify these flow directions, the characterization needs to describe the medium-specific migration directs that will exist after CCW placement, not merely conditions existing prior to placement.

To detect impacts from the environmental placement of CCW, a monitoring system must be able to detect and identify all contaminants migrating from the placement area. This requires there have been waste characterization that identifies all mobile concentrations of contaminants from the waste, seasonal variations in the mobile contaminants and their concentrations, and long-term changes in the population of mobile contaminants.

To detect impacts from the environmental placement of CCW, a monitoring location must be active when contaminants from the placement area are moving through the monitoring location. This requires there have been site characterization that is sufficient to project contaminant migration times to a point of observation. It also requires a monitoring program that remains in place long enough for contaminants to reach the monitoring point.

To detect impacts from the environmental placement of CCW, a monitoring system must be able to detect and identify contaminants mobilized by site leachates, whether or not the contaminant is itself released from the placement area. This requires there have been waste characterization and site characterization that is adequate to simulate the reactions between waste leachate and site soil and rock materials in contact with the leachate. For example, one presumed beneficial use for CCW is alkaline addition to areas that have long suffered from acid mine drainage. However, apparently obvious solutions can have unfortunate consequences. I have included another paper with this testimony that illustrates one example. My 2005 paper "Water Quality Impacts from Remediating Acid Mine Drainage with Alkaline Addition" explores the geochemistry that supports observations of arsenic contamination following the use of CCW as an alkaline addition, even when there is no evidence of excessive arsenic in the CCW leachate itself. Alkaline leachate from the CCW mobilizes previously-sequestered arsenic from on-site sediments.

Site Characterization

Adequate site characterization is seldom performed prior to approval for environmental placement of CCW. Depending upon the state, no site characterization may be required prior to some beneficial use placement. In other cases, something as simple as establishing the depth to water table prior to placement may be all that is required. Placement for beneficial use does not preclude negative environmental or health impacts, nor ensure that there is even a net improvement when benefits are weighed against negative impacts. Site characterization is as necessary at sites of environmental placement for beneficial use as for disposal.

When the placement activity includes site characterization, that characterization virtually always is of the conditions that exist prior to, not subsequent to waste placement. Seldom does CCW placement leave the hydrologic balance as it existed before placement. As a result, monitoring systems are designed to measure a flow system with no waste in it, not the one with waste present. This inadequacy is dramatically in evidence when considering placement in areas that have been mined or quarried. Coal mines and bedrock quarries typically entail huge dewatering programs. Coal ash placement, reclamation and bond release can occur decades before the mined areas reach full, equilibrium recharge, and during that time ground water is flowing into the void, not from it. The monitoring system, when there is one, is monitoring background water flowing toward the placement area, not water from the placement are and cannot possible convey information about the health or environmental risks associated with the eventual hydrologic system that will finally develop.

Site characterization seldom includes a characterization of the anticipated time-dependent variations of the site hydrogeology. This problem is very commonly observed for placement in coal mines. The Prides Creek Mine example in the previously cited "Environmental Concerns and Impacts of Power Plant Waste Placement in Mines" shows one case. Even when there is an intra-mine monitoring point that shows the strong temporal variability of water quality and ground water heads, there is no characterization to provide context for those changes.

Waste Characterization

This aspect of the various state-managed CCW programs is so weak as to be nearly meaningless in most states. Typically state programs use the results of the TCLP (toxic characteristics leaching procedure) or the SPLP (synthetic precipitation leaching procedure) as the predictor of the potential for placed CCW to impact health or the environment. This myopic misuse of laboratory index tests is probably the single greatest cause of the disconnect between the contamination that occurs from environmental placement of CCWs and what is promoted by advocates and regulators of the materials.

There is no justification for states to use these index laboratory tests as surrogates for determining likely field leachate for CCWs to be placed. These tests were not developed as predictors of field leachate, they are not designed to produce field leachate, and they have been repeatedly demonstrated incapable of doing so. The National Academies of Science understand this. The USEPA Science Advisory Board understands this. Yet, based upon the results of these inappropriate tests, multi-million ton masses of CCWs are allowed by some states to be placed without confinement and without monitoring in high risk hydrologic environments adjacent to private well users. And, the producers of the coal combustion and the regulators who approve it waste feign surprise or innocense when wells become contaminated.

If waste is to be placed in the environment, whether for disposal or beneficial use, complete and meaningful characterization is quite simply mandatory if the placement is to be protective. That testing, to be adequate for both instantaneous characterization of the waste and the design of the initial monitoring system, should include analyses of grain size and texture, elemental composition, chemical composition, mineralogy, rheology, hydrological properties, initial leachate compositions, and reactive potential with non-waste site soils and rock. Only with such characterization can any benefits be weighed against impacts and risks of placement. And only with such characterization can adequate monitoring be designed to confirm design predictions and measure site performance.

Finally, characterization of the CCW requires consideration of the time factor. The most abundant CCWs are highly reactive. They form in an environment that is completely out of equilibrium with the placement environment. Water is a solvent that carries dissolved contaminants away from the placement area and facilitates reaction with site soils, rock, and water. But it is also a major reactant with the wastes. Fly ash fresh from the burner is not the same material as fly ash that is quenched and sluiced to a pond. Nor is the fly ash that is dredged from the pond the same material as the fly ash in a pit five years after placement. CCWs evolve continually for years.

A test of ash fresh from the burner – whether for composition, mineralogy, texture, strength or leaching – will be different from that same ash after quenching and sluicing, which will be different after placement, which will be different 10, 20 and 50 years after placement. As the in-place ash evolves, so will the composition of the leachate from the ash. Contaminants that were sequestered in the young ash can become mobile the ash matures. Concentration of contaminants can rise, fall, and rise again, depending on the stage and sequences ash weathering. Eventually the glass component of some ashes can devitrify, producing late stage mobility of previously sequestered contaminants.

CCW characterization as performed today in the state programs virtually ignores the time factor and the recognition that ash will ultimately evolve to something quite far from its starting point. This is somewhat ironic with respect to CCW placement in coal mines. State mining regulators would laugh an operator from their offices who seriously proposed to use TCLP or SPLP to evaluate the acid producing potential of mine spoil. Yet, under state programs, those same regulators blithely allow those tests to predict the alkalinity that will be needed from the ash to neutralize delayed acid generation when CCW is used for alkaline addition. The 2002 evaluation I performed for Anker Energy and the West Virginia Highlands Conservancy, described in the attached paper "Assessment of the Anker Energy Corporation proposal for mining and reclamation, Upshur County, West Virginia," undertook a far more detailed evaluation of initial leaching characteristics. That evaluation, confirmed by an on-site pilot study, established that for the ash in question, the bulk of the ash's alkalinity would immediately flush from the placement area, leaving insufficient alkalinity available when acid mine drainage would be generated.

So long as waste characterization is driven by TCLP and SPLP results, there will be no reliability in the predicted results of CCW placement in the environment. And, based upon observed changes proposed and implemented in state programs, including increasing CCW masses in and approvals of unmonitored, unconfined placement for beneficial uses, it is apparent that direction is needed at the federal level.

THE FRAMEWORK

The successful construction of a new framework to address risks associated with CCW will require whole-hearted acceptance of the core element implicit in the question before this subcommittee, that it is appropriate for the federal government to step in to address the problems inherent in the management of these materials. Key elements of that framework are described in this section. Comparable elements of a framework applicable specifically to mine placements are described in the 2003 paper I produced for the USEPA and have attached with my testimony, "Developing Reasonable Rules for Coal Combustion Waste Placement in Mines. Why? When? Where? How?"

General Considerations

CCW is an industrial solid waste. Its placement must be in compliance with solid waste laws, clean air laws, and clean water laws. If it is placed or used in coal mines, placement also must be in compliance with state and federal surface mining and reclamation laws. State policy cannot be less protective than federal law.

Responsibility for the waste and any resulting damage remains with both the waste generator and the operator of the waste placement site.

Regulations must provide enforceable standards of both condition and performance, not merely discretionary guidelines. Oversight of the program must be by professionals trained and knowledgeable in waste disposal law, regulation, policy and practices. CCW placement site operators must demonstrate knowledge of, and the capability to fully implement waste disposal law, regulation, policy and practices.

Regulations must allow for public participation in the approval process, there must be the right of appeal, and cost recovery for successful appeal and citizen enforcement must be included.

Waste Characterization

Each CCW proposed for environmental placement shall be analyzed for grain size and texture, elemental and chemical composition, mineralogy, rheology, and hydrological properties. The constituent list will include all reasonably anticipated constituents of CCW and include tests for total radioactivity and radionuclides with environmental or health standards, and tests for polyaromatic hydrocarbons and other products of incomplete combustion of environmental and health concern.

When multiple CCWs are proposed for placement in a single location, the wastes shall be characterized individually, as above, and as a composited sample proportionate to the masses of the individual waste streams. This applies to both multiple waste streams from a single generator and waste streams from multiple generators.

Leachate Characterization

Prior to permit approval, the placement site operator will demonstrate to the extent possible the composition or limits on composition of the leachate(s) that will form at the site under the conditions of placement. This demonstration may include field testing, laboratory testing (sequential batch tests, column tests, etc.), computer modeling and/or other appropriate methodologies. The analyte list will be the same as for waste characterization.

For each placement area with different waste streams deposited, the placement site operator will install a monitoring well capable of sampling the leachate(s) that form in the field. Field leachate(s) will be sampled and analyzed for the same constituent list as for waste characterization.

Site Characterization

Site characterization will be comparable to that required for solid waste disposal facilities designed for wastes of comparable physical and chemical properties, and will use methodologies and protocols appropriate for solid waste disposal facilities.

Site geology will be characterized sufficiently to demonstrate the structure; bedrock stratigraphy; sediment, soil, spoil, fill, and waste distribution, composition, and texture; and geomorphology that will exist at and under the placement site(s) and in the adjacent areas.

Site hydrogeology will be characterized sufficiently to demonstrate the ground water and surface water systems and exchanges between them before, during and after CCW placement. The site characterization will include determining recharge areas, discharge areas, base flow contributions, hydraulic gradients, dominant flow paths, fluxes, velocities, travel times, physical properties (permeabilities, porosities, pore systems) for each material including CCW, water users and usable water resources, water chemistries, and the range of temporal variations typically experienced and likely to be experienced by any of these parameters. The description of this characterization will include a projection of the post-placement conditions.

Site characterization itself will be performed in a manner that will not be environmentally damaging to areas adjacent to or beneath the placement site(s).

Due to the highly transient stresses that will be imposed upon a placement facility during the construction, use and recovery, the site characterization should be continually updated through the life of the project as more data become available.

Fate and Transport of Leachate

Prior to issuance of the permit, the evolution of the chemistry of expected leachate(s) must be evaluated for each of the dominant flow paths as contact with ground water and migration through soil, and/or rock occurs. If the flow path involves the transport of leachate to a surface water system, the evaluation must include the evolution of the chemistry with respect to reactions with the mixing waters and the gases of the atmosphere. The evaluation will include major-, minor-, and trace-element compositions, and may be based upon field testing, laboratory testing (sequential batch tests, column tests, etc.), computer modeling and/or other appropriate methodologies.

If, after collection, actual field leachate differs significantly from the projected leachate(s), the evaluation(s) will again be performed using the field leachate composition(s).

The impacts of the leachate(s) on biota or on the uses of the water at receptors or compliance points will be evaluated, and the composition of the leachate(s) relative to applicable standards.

Monitoring

Prior to issuance of the permit, air, ground water, and surface water monitoring will be performed that is sufficient to document ambient air, ground water, and surface water quality; surface water quantity; and flux exchanges between ground and surface water for the range of temporal variations typically experienced at the placement site. Methodologies and protocols appropriate for waste disposal facilities will be used.

During the life of the placement operations, ongoing air monitoring of the placement work site and adjacent areas will be done for both dust and fugitive waste. Surface water discharges from the placement site will monitored for the full list of constituents used in characterization. Ground water will be monitored for both heads and chemistry, and surface water monitored for chemistry. The head data will be used to evaluate the validity of the site characterization and water chemistry will be used to verify that the CCW placement operation is not having negative impacts on downgradient or downstream water quality. Methodologies and protocols appropriate for waste disposal facilities will be used.

After placement is completed, ground and surface water monitoring will occur at locations and from wells capable of sampling leachate(s) from the placement site. Post-placement monitoring will continue until it is determined that leachate(s) have reached the wells, that site performance is as predicted in the permit, that the impacts and compositions at compliance points or receptors are within standards or are acceptable in the absence of standards, and are stable. Methodologies and protocols appropriate for waste disposal facilities will be used.

Compliance, Enforcement and Remediation

Compliance standards for each constituent of potential concern must be defined for surface water discharges, base flow discharges, placement-site air quality, fugitive dust off-site, fugitive waste off-site, and ground water.

Enforcement procedures must be defined and in place prior to permit issuance.

Remediation standards and procedures must be defined, and sufficient financial surety to perform necessary site, surface water, or ground water remediation must be demonstrated and maintained until monitoring is no longer required as provided above.

Isolation of Waste

If characterization and fate and transport analyses do not demonstrate that compliance will occur without barriers and or other containment procedures, the CCW placement cannot occur without extra measures to demonstrate compliance with performance requirements.

Informed Consent of Property Owners

Existing property owners must be advised of the following as part of obtaining consent for placement: a) the proposed activity is solid waste placement, b) the CCW will in all likelihood be or contain toxic forming material, c) the location(s), depth(s), and tonnages that may be placed of on the property, d) the source of the CCW(s), e) the composition of the CCW(s) and leachates, and f) that future buyers of the property have the right to disclosure of the CCW placement activity.

If CCW placement occurs at any surface mine, whether pre-law or post-law, the surveyed location, depth, quantity and character of the CCW shall be recorded with the deed for the property. This applies to state, corporate, private or abandoned mined lands.

Nomenclature and Rhetoric

Much time and fury is devoted to the nomenclature associated with the materials that remain after the combustion of coal with or without other fuels, far more time than is necessary or constructive.

> It's coal combustion (CC) waste. It's CC product. It's CC byproduct. It's CC residual. It's not "waste," because it can be reused. Until it is used in a product, it is a waste. It's pejorative to use "waste" and that makes it harder to convince people to reuse it. Euphemistic phraseology lowers the perception of the need for protection. Dumping of these materials should be managed like the disposal of any other waste. It's not being disposed, it's being beneficially used. *Ad nauseam.*

There is a method to the verbal madness, of course. If one defines the vocabulary, one controls the debate. It's why trade organization employees monitor and control even the text of Wikipedia entries on combustion wastes.

Although the policy debate is influenced by the vocabulary, the reality and the science are not. Filling an open, unlined pit in Indiana with fly ash, while calling it landfill disposal, ruined an aquifer and created a SuperFund site. Had it been called beneficial use, it would have still ruined the aquifer and created the SuperFund site. There would be no lower environmental and health risks were it called coal combustion product instead of coal combustion waste. The ill-chosen placement methodology of inappropriate CCW created the problem, not the nomenclature, and changing the labeling does not change the chemistry or the hydrogeology a whit.

However, increased sophistication in language management has changed CCW regulation in Indiana. Because the Pines ash would pass the TCLP/SPLP characterization criteria, placement of Pines-like ash in a Pines-like pit today can be called beneficial use, structural fill. For beneficial-use placement as structural fill, Indiana doesn't require ground water monitoring, the kind of monitoring that ultimately allowed the citizens Pines to document their contamination. The program improvement for industry is that industry can claim a higher rate of "reuse" of this CCW for exactly the same placement practices. The program improvement for Indiana is that it needn't see a problem. And neither industry nor Indiana has to deal with a SuperFund site. If the next Pines is to be avoided, its citizens need help from Washington.

Don't mistake these comments as a criticism focused on Indiana or of pit-filling. Recently, in Virginia, several millions of tons of a CCW source that created contamination problems at a controlled, monitored, on-site landfill was approved for the "beneficial use" of sculpting rolling terrain for a golf course. The placement is without containment, without leachate collection, and without the monitoring of a disposal facility than could detect a problem. The site characterization consisted of determining the elevation of the water table pre-project, not after completion of the placement. The waste characterization was by TCLP and/or SPLP. It is in compliance with Virginia regulations. Media and citizen concern over the disconnect of problems at a permitted waste disposal facility and open placement in a neighborhood led to this spring to testing of residential wells adjacent to the placement area. The initial, limited testing by the city, not the Commonwealth, identified problematic concentrations in some wells of boron, a common contaminant in fly ash. Further evaluations are continuing by citizens and by the city.

Time and further evaluation will tell if the golf course is an early-stage Pines. But, the evolution so far is eerily similar to that at Pines. Local investigation finds a water problem with citizens' drinking water. Regulators assure that the placement in the neighborhood of a waste with a history of problems was done in compliance with their regulations. "But, what about the water?" "The waste placement complies." Something isn't right, and needs to be fixed.

In an absence of meaningful direction and oversight at the federal level, state regulation has entered a race to the bottom with respect to regulatory control over placement of CCW. The definitions of "beneficial use" are expanding and the criteria of a waste to qualify are relaxing. There is a concomitant relaxation of management controls, waste and site characterization, and monitoring. The cycle creates the statistical illusion of increased "reuse" while setting up longterm environmental and health problems in state after state; in mines, gravel pits, quarries, or simply fills. Increasing, the public is blind to the development of problems. And, as one state relaxes the controls yet further, others competitively follow.

As the acceptance of the beneficial use approach deepens at the state level, documentation of the problems becomes increasingly difficult because there are no monitoring data. One actually hears the argument in favor of beneficial use that there are no problems seen at beneficial-use sites, unlike waste disposal sites. My grandson, by age three, knew that covering his eyes didn't make spilled juice go away. It is sophistry to argue that no evidence of impacts, as the result of not looking for impacts, is affirmative evidence of no impacts. Yet that is just what some proponents of environmental placement of CCWs suggest.

There is another problem developing out of the CCW management approach that, while unrelated directly to CCW, will be impacted by the actions of this committee. Until credibility is brought to the regulation of coal combustion materials, there will be increasing collateral damage as well. State regulators are being approached by industry to implement the beneficial-use approach to other waste stream, particularly with respect to the misuse of the result of TCLP and SPLP. Functionally, the argument becomes, "If I can control the chemistry of a handful my waste for the eighteen hours of your lab test, I should be allowed to place my waste in the environment, without containment and without monitoring, just like you allow for CCW, for a beneficial use." The argument is even being extended by one Illinois company for delisting of at least one listed RCRA hazardous waste.

CONCLUSION

A rose by any other name still has its thorns. Labeling an environmental placement of CCW a beneficial use does not reduce damage that may be done or the risk to health and the environment. If a CCW has a legitimate beneficial impact, one that can demonstrated and quantified, do so, and analyze the entire costs and risks of the placement, and compare that with quantified benefits. Maybe for a particular placement, the benefit is projected to exceed the impacts and increased risk to health and the environment. Even when so, stewardship of the placement is critical to verify nothing was done that wasn't projected. But that approach is not the approach today, and the shift will have to come with federal involvement.

As a society, we used federal action to reduce the health and environmental risks of the physical and chemical rain from the stacks decades ago. For most of us, the air improved, and with it, the environment and our health. But the toxins don't go away; we just capture them. Just as a federal framework was needed then to guide states in addressing risks to health and the environment by dispersing these materials, it is needed again to address the risks from the same material, now accumulated instead of dispersed.

Again, thank you.

In: Coal Combustion Waste: Management and Beneficial Uses ISBN: 978-1-61728-962-0
Editor: Daniel D. Lowell

Chapter 14

WRITTEN TESTIMONY OF MARY A. FOX, PHD, MPH, ASSISTANT PROFESSOR, JOHNS HOPKINS BLOOMBERG SCHOOL OF PUBLIC HEALTH, BEFORE THE SUBCOMMITTEE ON ENERGY AND MINERAL RESOURCES, HEARING ON "HOW SHOULD THE FEDERAL GOVERNMENT ADDRESS THE HEALTH AND ENVIRONMENTAL RISKS OF COAL COMBUSTION WASTE?"

INTRODUCTION

I thank you for the opportunity to testify today concerning the health effects of exposure to coal combustion waste. I am Dr. Mary Fox, Assistant Professor in the Department of Health Policy and Management in the Johns Hopkins Bloomberg School of Public Health. I am a risk assessor with doctoral training in toxicology, epidemiology and environmental health policy. I am a core faculty member of the Hopkins Risk Sciences and Public Policy Institute where I teach the methods of quantitative risk assessment. In my research I evaluate the health risks of exposure to multiple chemical mixtures.

My testimony focuses on the health effects associated with exposure to coal combustion waste and assessing the public health risks of such exposures.

BACKGROUND

According to a recent report from the National Research Council, coal combustion waste includes several waste streams produced at coal-fired facilities, for example, bottom ash and boiler slag from the furnace, and fly ash and flue gas desulfurization material collected by pollution control devices (NRC 2006). The amount produced annually in the US exceeds 120 million tons or enough to fill a million railroad coal cars (NRC 2006). Coal combustion waste has numerous inorganic constituents, many of which are associated with health effects in studies of animal or human exposures. Exposures to human populations may occur depending on methods of coal combustion waste disposal. A summary of health effects information for coal combustion waste constituents following studies of oral (ingestion) exposures is provided below.

From a public health perspective it is interesting to note that the current concerns about coal combustion waste disposal are in part a result of regulatory success at protecting air quality. Two of the waste streams that contribute to the total production of coal combustion waste are from pollution control technologies in place to maintain clean air. Our efforts to minimize air emissions have resulted in a shifting of toxic constituents to another less well-regulated waste stream with potential to release the toxins into other environmental media.

Evaluating Potential Health Risks from Exposure to Coal Combustion Waste

Methods of Coal Combustion Waste Disposal and Potential for Human Exposure

Several methods of coal combustion waste disposal were identified by the National Research Council committee including placement in lined or unlined landfills, placement in lined or unlined surface impoundments, use in engineered products such as cement, placement or use in coal mines (NRC 2006). If the coal combustion waste is in contact with surface water or groundwater there is potential for the waste to be mobilized into the surrounding environment by leaching or runoff. During transport or placement (dumping) coal combustion waste may be entrained in air. Humans may come into contact or be exposed to coal combustion waste that has been mobilized into the environment from a disposal site. For example, if coal combustion waste leachate is in groundwater it may reach drinking water wells. Coal combustion waste entrained in air may be inhaled, may settle on soil or be transported into buildings through air transfer or on shoes or clothes.

Management of coal combustion waste is a national issue that affects communities around the country where disposal sites are located. Not far from here in Anne Arundel County, Maryland, coal combustion waste has been disposed of in a sand and gravel pit. The county health department has sampled the drinking water wells of nearby residents finding concentrations of aluminum, arsenic, beryllium, cadmium, lead, manganese, and thallium at levels above primary and secondary drinking water standards in some wells (Phillips 2007). It appears that coal combustion waste buried in the former sand and gravel pit is leaching into groundwater.

Health Effects Information on Constituents of Coal Combustion Waste

Health effects information is available for the majority of coal combustion waste constituents. See Table 1. The types and severity of the health effects range from benign and cosmetic effects to changes in organ or system function to cancer. Several coal combustion waste constituents share a common type of toxicity or target organ or system. Three coal combustion waste constituents have neurological effects (aluminum, lead, manganese); three (barium, cadmium, mercury) have effects on the kidney; three have a variety of effects on blood (cobalt, thallium, zinc); two have effects on the gastrointestinal system (beryllium and copper). If exposures to these mixtures occur, there is a greater chance of increased risk to health.

Table 1. Health Effects of Coal Combustion Waste (CCW) Constituents

CCW Constituent	Health Effect(s) of Concern (Exposure by Ingestion)	Information Source
Aluminum	Neurological	ATSDR 2007
Antimony	Longevity, changes in blood glucose and cholesterol	EPA IRIS
Arsenic	Cancer, hyperpigmentation, keratosis of skin	EPA IRIS
Barium	Nephropathy	EPA IRIS
Beryllium	Gastrointestinal	EPA IRIS
Boron	Decreased fetal weight	EPA IRIS
Cadmium	Significant proteinuria	EPA IRIS
Chromium (III)	No effects observed	EPA IRIS
Chromium (VI)	No effects observed	EPA IRIS
Cobalt	Blood	ATSDR 2007
Copper	Gastrointestinal	ATSDR 2007
Fluorine	Cosmetic fluorosis of teeth	EPA IRIS
Iron	NA	NA
Lead	Neurological	CDC 2005
Manganese	Neurological	EPA IRIS
Mercury	Kidney	ATSDR 2007
Molybdenum	Increased uric acid levels	EPA IRIS
Nickel	Decreased body and organ weight	EPA IRIS
Potassium	NA	NA
Selenium	Selenosis – hair and nail loss	EPA IRIS
Silver	Argyria - benign skin pigmentation	EPA IRIS
Strontium	Bone growth and mineralization	EPA IRIS
Thallium	Change in blood chemistry	EPA IRIS
Vanadium	Decreased hair cystine	EPA IRIS
Zinc	Decreased red blood cell copper and enzyme activity	EPA IRIS

Abbreviations: ATSDR, Agency for Toxic Substances and Disease Registry; CCW, coal combustion waste; EPA, Environmental Protection Agency; IRIS, Integrated Risk Information System; NA, not available.

The health effect information for coal combustion waste constituents in Table 1 was gathered from the Centers for Disease Control and Prevention (CDC), the Agency for Toxic Substances and Disease Registry (ATSDR) and the U.S. Environmental Protection Agency Integrated Risk Information System (IRIS). The health effects information listed comes from studies of exposure by ingestion. The listing of coal combustion waste constituents was developed from the National Research Council 2006 report "Managing Coal Combustion Residues in Mines".

Assessing Risks to Human Health

Environmental public health agencies such as the US Environmental Protection Agency routinely use human health risk assessment to evaluate health impacts of exposure to contaminated environmental media such as air and drinking water. Human health risk assessment is a systematic process that combines available data on the contaminant of concern as described in the National Research Council report "Risk Assessment in the Federal Government: Managing the Process" (NRC 1983). The four basic steps of a human health risk assessment are hazard identification, dose-response assessment, exposure assessment and risk characterization. Hazard identification summarizes information on the health effects related to exposure to the contaminant of concern. (As presented in Table 1, hazard information is known for the majority of coal combustion waste constituents.) Dose-response data are developed from research studies and describe the quantitative relationship between exposures and changes in rates of diseases, or other health effects such as organ function changes. Dose-response data are available for the majority of coal combustion waste constituents presented in Table 1. The magnitude, duration and amount of contact the individual or population of concern has with the contaminant of concern will be described in the exposure assessment. The nature of exposure to coal combustion waste will be highly variable depending on conditions at the site of disposal. The risk characterization combines the exposure and dose-response data to evaluate the likelihood of increased health risk.

Human health risk assessment methods are available to evaluate multiple chemical exposures (EPA 2000). Coal combustion waste is a complex mixture of constituents. Risk assessment methods for multiple chemical exposures will be essential to evaluating health risks of exposure to coal combustion waste.

Three of the four common coal combustion waste management practices (landfill, surface impoundment, use in or reclamation of mines) result in localized disposal. Communities surrounding such disposal sites are typically small. Proximity to the coal combustion waste disposal site will likely spur interest in evaluating community health. Unfortunately, systematic health effects research in any one small community will have limited statistical power to detect changes in health outcomes.

Reducing Risks to Human Health

Risks to human health are increased if people are exposed to coal combustion waste. The tremendous volume of this waste generated and disposed of each year in communities throughout the country represents an enormous public health challenge. People are exposed if coal combustion waste is dispersed into the broader environment by runoff, leaching or entrainment in air. Dispersal of coal combustion waste into the broader environment will be reduced or eliminated by disposal practices that contain the waste away from contact with ambient air, surface water and groundwater. Human health risks are reduced or eliminated if human exposure is reduced or eliminated.

CONCLUSIONS

Coal combustion waste is a mixture of well-recognized substances. The approach to evaluating exposures to coal combustion waste should acknowledge potential interactions among the constituents in the body. Methods are available to assess health risks from exposure to mixtures of chemical substances, however, current regulatory strategies were not designed to control such mixture exposures. Coal combustion waste disposal practices must be improved to ensure population exposures are controlled through appropriate long-term containment and management.

Main points:

- Large volumes of coal combustion waste are produced and disposed of in the U.S. every year.
- Coal combustion waste is a complex mixture that can become mobilized in the environment, depending on disposal methods used.
- People are exposed through multiple means including inhalation, direct contact, and ingestion. Exposures may occur indoors and outdoors.
- Current approaches to evaluating health risks are limited and may underestimate the true risks to exposed communities.
- Health effects of exposure will be underestimated unless the potential cumulative impacts of the multiple toxic components of the mixture are considered together.
- Prevention of exposure through better management of the waste is ultimately the most sound public health approach.

Thank you very much for this opportunity to address the Subcommittee.

REFERENCES

Agency for Toxic Substances & Disease Registry. (2007). *Minimum Risk Levels*. Available at: http://www.atsdr.cdc.gov/mrls/index.html [accessed May 23, 2008].

Centers for Disease Control and Prevention (2005). *Preventing Lead Poisoning in Young Children*. Atlanta: CDC.

Environmental Protection Agency. (2000). *Supplementary Guidance for Conducting Health Risk Assessment of Chemical Mixtures*. EPA/630/R-00/002. Risk Assessment Forum, Washington, DC.

Environmental Protection Agency. (2008). *Integrated Risk Information System*. Available at: http://cfpub.epa.gov/ncea/iris [accessed May 23, 2008].

National Research Council Committee on the Institutional Means for Assessment of Risks to Public Health. (1983). Risk Assessment in the Federal Government: *Managing the Process*. Washington: National Academy Press.

National Research Council Committee on Mine Placement of Coal Combustion Wastes. (2006). *Managing Coal Combustion Residues in Mines*. Washington: The National Academies Press.

Phillips, F. (2007). *Impacts of Fly Ash on Groundwater in Anne Arundel County, Maryland.* Available at: www.mde.state.md.us/assets [accessed May 23, 2008].

In: Coal Combustion Waste: Management and Beneficial Uses ISBN: 978-1-61728-962-0
Editor: Daniel D. Lowell

Chapter 15

TESTIMONY OF LISA EVANS, PROJECT ATTORNEY, EARTHJUSTICE, BEFORE THE SUBCOMMITTEE ON ENERGY AND MINERAL RESOURCES, HEARING ON "HOW SHOULD THE FEDERAL GOVERNMENT ADDRESS THE HEALTH AND ENVIRONMENTAL RISKS OF COAL COMBUSTION WASTE?"

Chairman Costa and Members of the Subcommittee, thank you for holding this hearing to consider the federal government's role in addressing the health and environmental risks of coal combustion waste. When mismanaged, coal combustion waste damages aquatic ecosystems, poisons drinking water and threatens the health of Americans nationwide. One of the dangers posed by coal combustion waste is disposal in coal mines, a practice that threatens the already heavily impacted communities and natural resources of our nation's coal mining regions.

I am Lisa Evans, an attorney for Earthjustice, a national non-profit, public interest law firm founded in 1971 as the Sierra Club Legal Defense Fund. Earthjustice represents, without charge, hundreds of public interest clients in order to reduce water and air pollution, prevent toxic contamination, safeguard public lands, and preserve endangered species. My area of expertise is hazardous and solid waste law. I have worked previously as an Assistant Regional Counsel for the Environmental Protection Agency enforcing federal hazardous waste law and providing oversight of state programs. I appreciate the opportunity to testify this morning.

The question before this subcommittee, how the federal government should address the risks of coal combustion waste, has a straightforward answer. Simply stated, the U.S. Environmental Protection Agency (EPA) must do what it committed to do in its final *Regulatory Determination on Wastes from the Combustion of Fossil Fuels*, published 8 years ago.[1] In that determination, mandated by Congress in 1980, EPA concluded that federal standards for the disposal of coal combustion waste under the Resource Conservation and Recovery Act (RCRA) and/or the Surface Mining Control and Reclamation Act (SMCRA) are required to protect health and the environment. EPA's commitment to set minimum federal disposal standards extended to coal ash disposed in landfills, lagoons and mines. Yet eight years later, and 25 years after Congress required this determination, EPA's commitment remains an entirely empty promise.

The failure to fulfill this commitment is wholly unjustified, particularly in light of the substantial research that has already been completed by both EPA and the National Academies of Science (NAS). Preceding EPA's 2000 determination, EPA complied (albeit 16 years late) with a congressional mandate under RCRA to study the risks posed by coal combustion waste, solicit public comment, hold a public hearing, and publish a Report to Congress.[2] As a result, there is a robust record documenting the risks posed by coal ash and the damage that has occurred throughout the country as a result of its mismanagement. Further supplementing the record, EPA published in August 2007 a Notice of Data Availability that included additional documentation of the risks posed by coal combustion waste including a draft *Human Health and Ecological Risk Assessment* and a *Coal Combustion Waste Damage Case Assessment.* Lastly, EPA's Office of Research and Development has published a series of documents detailing the increasing toxicity of coal combustion waste, including *Characterization of Mercury-Enriched Coal Combustion Residues from Electric Utilities Using Enhanced Sorbents for Mercury Control.*

Secondly, in 2004, Representative Nick Rahall introduced legislation requiring the NAS to study the impact of coal ash placement in mines and to recommend what federal action, if any, should be taken to control this burgeoning practice. In March 2006, the NAS published a report, *Managing Coal Combustion Residues in Mines*, that concluded unequivocally that enforceable federal standards be established to protect ecological and human health. The NAS recommended that EPA and the U.S. Office of Surface Mining (O SM) work together to promulgate federal standards under RCRA, SMCRA or a combination of both statutes.

It is now two years since the publication of the NAS report, 8 years after EPA's final regulatory determination, and 28 years since Congress first asked EPA to study the question. While the federal agencies have failed to act, the need to resolve this question has become increasingly urgent. When one considers the escalating number of sites polluted by coal combustion waste, the documented increase in the toxicity of coal ash, the increase in U.S. coal use, the accompanying increase in the volume of waste, and the trends in mismanagement, the path is clear. Flying blind without federal rules that ensure safe disposal of the largest industrial waste in the country is nothing if not foolish, dangerous, and contrary to statutory mandates and clear Congressional intent.

EPA and OSM are fiddling while ash from burning coal poisons our water and sickens our communities. Inadequate state laws offer scant protection. Federal environmental statutes dictate that EPA and OSM must do what they promised to do and what they have been directed to do -- promulgate enforceable minimum federal standards to protect health and the environment nationwide from the risks posed by mismanagement of coal combustion waste.

THE NATURE OF THE THREAT FROM COAL COMBUSTION WASTE

1. The Volume of Waste is Immense

Burning coal produces over 129 million tons *each year* of coal combustion waste in the U.S. This is the equivalent of a train of boxcars stretching from Washington, D.C. to

Melbourne, Australia.[3] Coal combustion waste (CCW) is largely made up of ash and other unburned materials that remain after coal is burned in a power plant to generate electricity. These industrial wastes include the particles captured by pollution control devices installed to prevent air emissions of particulate matter (soot) and other gaseous pollutants from the smokestack. In addition to burning coal, some power plants mix coal with other fuels and wastes, including a wide range of toxic or otherwise hazardous chemicals, such as the residue from shredded cars (a potential source of PCBs), oil combustion waste (often high in vanadium), railroad ties, plastics, tire-derived fuel and other materials.[4]

As demand for electricity increases and regulations to reduce air emissions from power plants are enforced, the amount of CCW is expected to increase. By 2015, the quantity of CCW generated per year is estimated to exceed 170 million tons. (See Figure 1) In addition, the Energy Information Administration (EIA)'s *2007 Annual Energy Outlook* indicates that electricity production from coal is projected to increase almost 25 percent by 2020 and 64% by 2030.[5] Production of CCW will increase proportionally.

CCW is significantly different from coal itself. As coal is burned, its volume is reduced by two thirds to four fifths, concentrating metals and other minerals that remain in the ash. Elements such as chlorine, zinc, copper, arsenic, selenium, mercury, and numerous other dangerously toxic contaminants are found in much higher concentrations on a per volume basis in the ash compared to the coal. These wastes are poisonous and can cause cancer or damage the nervous systems and other organs, especially in children. The thousands of tons of chemicals disposed of in CCW each year dwarf other industrial waste streams. (See Figure 2) Table 1 below indicates some of the contaminants commonly found in CCW and their human health effects.

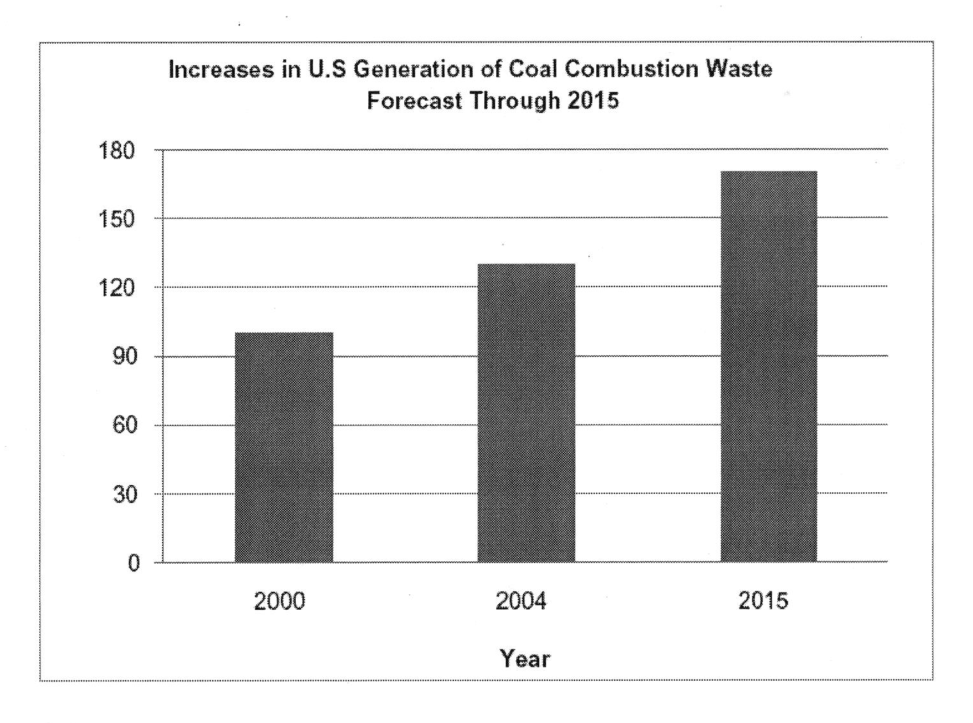

Figure 1

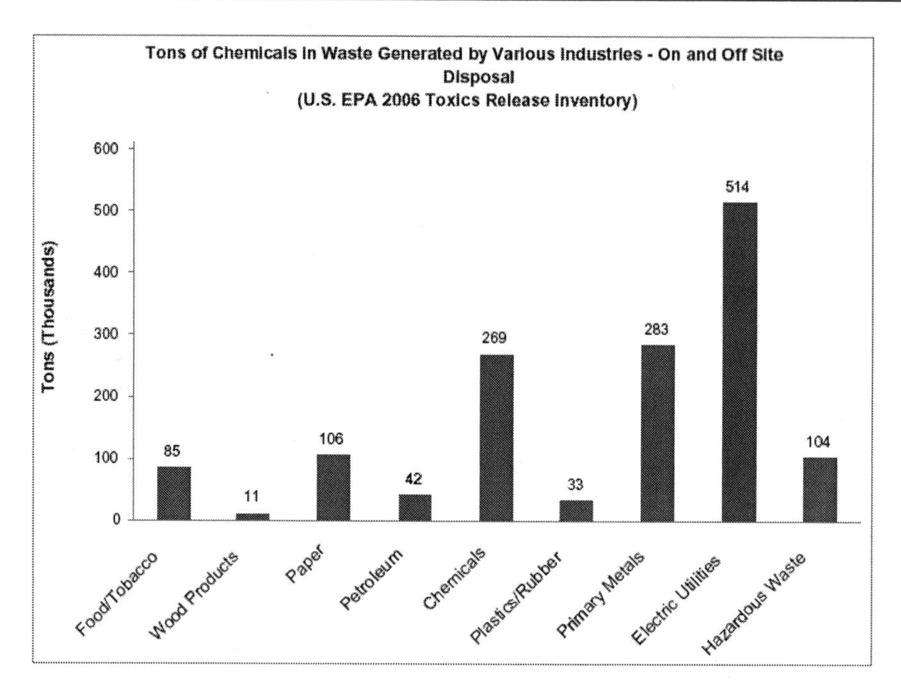

Figure 2

Table 1. Human Health Effects of Coal Combustion Waste Pollutants

Aluminum	Lung disease, developmental problems
Antimony	Eye irritation, heart damage, lung problems
Arsenic	Multiple types of cancer, darkening of skin, hand warts
Barium	Gastrointestinal problems, muscle weakness, heart problems
Beryllium	Lung cancer, pneumonia, respiratory problems
Boron	Reproductive problems, gastrointestinal illness
Cadmium	Lung disease, kidney disease, cancer
Chromium	Cancer, ulcers and other stomach problems
Chlorine	Respiratory distress
Cobalt	Lung/heart/liver/kidney problems, dermatitis
Lead	Decreases in IQ, nervous system, developmental and behavioral problems
Manganese	Nervous system, muscle problems, mental problems
Mercury	Cognitive deficits, developmental delays, behavioral problems
Molybdenum	Mineral imbalance, anemia, developmental problems
Nickel	Cancer, lung problems, allergic reactions
Selenium	Birth defects, impaired bone growth in children
Thallium	Birth defects, nervous system/reproductive problems
Vanadium	Birth defects, lung/throat/eye problems
Zinc	Gastrointestinal effects, reproductive problems

Source: ATSDR ToxFAQs, available at www.atsdr.cdc.gov/toxfaq.html

2. Better Air Pollution Controls Make CCW More Toxic

CCW is becoming increasingly toxic. As air pollution control regulations are implemented under the Clean Air Act, more particulates and metals are captured in the ash instead of being emitted from the smokestack. In a 2006 report on CCW, EPA found that when activated carbon injection was added to a coal-fired boiler to capture mercury, the resulting waste leached selenium and arsenic at levels sufficient to classify the waste as "hazardous" under RCRA.[6] Specifically, EPA found that arsenic leached (dissolved) from the CCW at levels as high as 100 times its maximum contaminant level (MCL) for drinking water, and selenium leached at levels up to 200 times its MCL.[7]

In a follow-up study that is currently underway by EPA's Office of Research and Development, EPA tested the leaching characteristics of CCW from a power plant employing both mercury controls and a wet scrubber for sulfur dioxide control. EPA found that CCW from a plant with a wet scrubber leached numerous additional toxic metals at levels significantly higher than their MCLs.[8] EPA found that the CCW leached arsenic, thallium, boron, and barium above RCRA's hazardous waste threshold (100 times the MCL). The CCW also leached levels of antimony, cadmium, chromium, lead, mercury, molybdenum and selenium in quantities sufficient to contaminate drinking water and harm aquatic life.

EPA's own analyses of how CCW behaves in unlined disposal sites predict that some metals will migrate and contaminate nearby groundwater to conditions extremely dangerous to people. In 2007, EPA published a draft *Human Health and Ecological Risk Assessment* that found extremely high risks to human health from the disposal of coal ash in waste ponds and landfills. According to EPA, the excess cancer risk for children drinking groundwater contaminated with arsenic from CCW disposal in unlined ash ponds is estimated to be as high as nine in a thousand - 900 times higher than EPA's own goal of reducing cancer risks to less than one-in-one hundred thousand individuals. Figure 3 compares EPA's findings on the cancer risk from arsenic in coal ash disposed in waste ponds to several other cancer risks, along with the highest level of cancer risk that EPA finds acceptable under current regulatory goals.

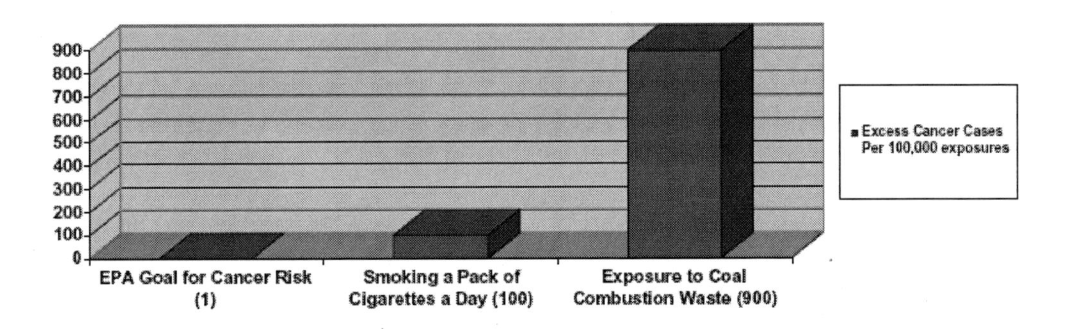

Figure 3

Clearly, as new technologies are mandated to filter air pollutants from power plants, cleaning the air we breathe of smog, soot and other harmful pollution, the quantity of dangerous chemicals in the ash increases. Without adequate safeguards, the chemicals that have harmed human health for years as air pollutants- mercury, arsenic, lead and thallium-will now reach us through drinking water supplies. Given the documented tendency of CCW to leach metals at highly toxic levels, there is clearly the need for scrutiny of current disposal practices.

3.CCW Causes Documented Damage to Human Health and the Environment

The absence of national disposal standards has resulted in environmental damage at disposal sites throughout the country. In fact, scientists have documented such damage for decades. Impacts include the leaching of toxic substances into soil, drinking water, lakes and streams; damage to plant and animal communities; and accumulation of toxins in the food chain.[9,10] According to EPA's latest *Damage Case Assessment for Coal Combustion Waste* published in 2007, EPA recognizes 67 contaminated sites in 23 states where CCW has polluted groundwater or surface water. EPA admits that this is just the tip of the iceberg, because most CCW disposal sites in the U.S. are not adequately monitored.

Low-income communities and people of color shoulder a disproportionate share of the health risks from these wastes. The poverty rate of people living within one mile of coal combustion waste disposal sites is twice as high as the national average, and the percentage of non-white populations within one mile is 30 percent higher than the national average. Similarly high poverty rates are found in 118 of the 120 coal-producing counties, where CCW increasingly are being disposed of in unlined, under-regulated mines, often directly into groundwater.

Documented damage from CCW includes:

- Public and private drinking water contaminated by CCW in at least 8 states, including Wisconsin, Illinois, Indiana, New Mexico, Pennsylvania, North Dakota, Georgia and Maryland.[11]
- Hundreds of cattle and sheep killed and many families sickened in northern New Mexico by ingesting water poisoned by CCW.[12]
- Fish consumption advisories issued in Texas and North Carolina for water bodies contaminated with selenium from CCW disposal sites and entire fish populations destroyed.[13,14]
- Documented developmental, physiological, metabolic, and behavioral abnormalities and infertility in nearly 25 species of amphibians and reptiles inhabiting wetlands contaminated by CCW in South Carolina.[15]

Unfortunately, new CCW-contaminated sites are being uncovered with disturbing frequency. One need only pick up the *Washington Post*, *Baltimore Sun* or *Virginian-Pilot* over the last few months to grasp the national crisis. Evidence of poisoned water has recently surfaced in Baltimore, Charles County, Virginia Beach, and across the country in Illinois, Indiana, and Montana.

The following sites are illustrative:

- **Gambrils Fly Ash Site, Anne Arundel County, Maryland** where 3.8 million tons of ash were dumped in unlined gravel pits contaminating drinking water wells with arsenic, lead, cadmium, nickel, radium and thallium as high as 4 times the drinking water standard.

- **Faulkner Landfill, Charles County, Maryland** where leaching coal ash is contaminating a wetland with selenium and cadmium at levels high enough to kill any animal life, The Smithsonian Institution has called the affected wetlands, Zekiah Swamp, one of the most ecologically important areas on the East Coast.
- **Battlefield Golf Course, Chesapeake, Virginia** where developers used 1.5 million tons of fly ash to build a golf course over a shallow aquifer. Although the course was just completed this winter, wells are already starting to show elevated boron. Investigation into the cause of the pollution has just begun. Residential drinking water wells are in close vicinity to the unlined, uncapped site.
- **PPL Montana Power Plant, Colstrip, Montana,** the second largest coal-fired power plant west of the Mississippi, where leaking unlined coal ash ponds contaminated residential wells with high levels of metals, boron and sulfate. Five companies agreed in May 2008 to pay $25 million to settle a groundwater contamination lawsuit brought by residents.
- **Gibson Generating Station, Gibson County, Indiana** where enormous ash ponds are exposing threatened species to dangerous levels of selenium and where the power company supplies residents with bottled water because their wells are contaminated with boron.

These injuries to human lives and the environment are entirely avoidable. Yet damage will continue to occur at site after site in the absence of minimum federal standards. As you read this testimony, approximately 1000 tons of ash is disposed daily into a New Mexico mine, although the mine continues to leach toxic levels of sulfate into scarce New Mexico waters. Constellation Energy, the company that poisoned the water in Gambrills, Maryland and paid a million dollar fine for that offense, is today seeking to dump its ash into another unlined Maryland quarry because there are no state laws prohibiting the dumping. And currently there is a permit pending in Pennsylvania that seeks to create the largest unlined coal ash dump in the U.S in a surface coal mine without any requirements for sufficient monitoring, waste or site characterization, cleanup standards, or bonds for cleanup. The damage that will result from these acts is not inevitable. It is within this subcommittee's power to require federal agencies to do their job to protect health and the environment from this toxic waste.

4. CCW is Disposed in Coal Mines without Safeguards

Each year, approximately 25 million tons of CCW, nearly 20% of total CCW generation, are placed in active and abandoned coal mines without basic safeguards to protect health and water resources. Under pressure from electric utilities, many states have wrongly defined the dumping of CCW in coal mines as a "beneficial use" and exempted the practice from all solid waste regulations.[16] Consequently, enormous quantities of CCW are being dumped directly into groundwater without any monitoring or clean up requirements.

The *laissez faire* regulatory approach of many states to CCW minefilling maximizes the risk of contamination. Mining breaks up solid rock layers into small pieces, called spoil. Compared to the flow through undisturbed rock, water easily and quickly infiltrates spoil that has been dumped back into the mined out pits. Fractures from blasting and excavation become underground channels that allow groundwater to flow rapidly offsite. Because mines usually excavate aquifers (underground sources of water), the spoil fills up with groundwater. Unlike engineered landfills, which are lined with impervious membranes (clay or synthetic) and above water tables by law, ash dumped into mine pits continually leaches its toxic metals and other contaminants into the water that flows through and eventually leaves the mine.

In fact, serious contamination has been documented at numerous mine sites across the country where CCW has been disposed. In a multi-year study of 15 coal ash minefills in Pennsylvania, researchers found that CCW made the water quality worse at 10 of the 15 mines.[17] At five of the sites, there was not enough monitoring data to determine whether adverse impacts were caused by the CCW. A review of the permits revealed that:

- Levels of contaminants, including manganese, aluminum, arsenic, lead, selenium, cadmium, chromium, nickel, sulfate and chloride, increased in groundwater and/or surface water after CCW was disposed in the mines.
- Contaminants increased from background concentrations (measured after mining) to levels hundreds to thousands of times federal drinking water standards.
- Pollution was found downstream from CCW disposal areas and sometimes well outside the boundary of the mines.

Even though the placement of coal ash in coal mines is often touted as a “beneficial use” for the purpose of treating acid mine drainage, the facts show that minefilling is not an effective solution. While the CCW remediated acid mine drainage *temporarily* in a few of the mines studied, in two thirds of the mines, the introduction of CCW resulted in more severe, long-term contamination than had existed at these sites from the mining operation itself. Furthermore, the stakes are high if contamination occurs. As a practical matter, dumping large quantities of CCW directly into water tables at highly fractured sites under massive quantities of mine overburden makes the prospect of cleaning up contamination far more daunting than halting leakages from conventional landfills and ash ponds.

5. States Fail to Provide Adequate Regulation of CCW Disposal

With no minimum federal standards, the states have been free to regulate as they please, or more often, abstain from effective regulation altogether. If one compares how EPA regulates the disposal of ordinary household trash with its hands-off approach to CCW, the results defy logic. While newspapers, soda cans and banana peels under no circumstances qualify as RCRA hazardous waste, EPA has established detailed federal disposal standards for the landfills that contain them.[18] Household trash cannot be dumped in a mine without violating federal law, but in most states battleship quantities of metal-laden ash can be dumped with relative impunity. EPA has regulations governing all aspects of the disposal of household trash in landfills including performance standards, siting restrictions, monitoring, closure requirements, bonding, and post-closure care.[19] These regulations, promulgated under subtitle D of RCRA, are enforceable by states and citizens against any owner or operator of a landfill in violation of the standards. Furthermore, RCRA requires that state solid waste programs promulgate equivalent (or more stringent) regulations in order to maintain authorization.[20] Yet EPA has no such regulations for the disposal of toxic ash that exceeds *hazardous waste* levels for toxic metals. The result is an inconsistent patchwork of largely inadequate state regulation.

The utility industry, as well as some states, claim that the states are doing a good job of regulating coal ash despite the absence of federal standards. The fact that EPA admits at least 67 sites in 23 states have been contaminated by CCW indicates that the opposite is true. A survey of state laws governing CCW disposal in landfills and surface impoundments shows that state regulations fall short of requiring measures that would adequately protect human health and the environment. Earthjustice, along with several other environmental organizations, submitted analyses of the laws and regulations of 20 states in response to EPA's Notice of Data Availability in February 2008. Our state survey is too voluminous to repeat in this testimony, but the analyses show definitively that state solid waste programs do not provide consistent and adequate safeguards sufficient to protect human health and the environment from CCW. Most states failed to require the basic safeguards essential for waste management, including liners, leachate collection systems, groundwater monitoring, corrective action (cleanup), closure and post-closure care.

In fact, the gaps are shocking. Among the top 15 CCW generating states, which represent 74% of U.S. CCW generation, *only one state* requires all CCW lagoons (surface impoundments) to be lined and *only one state* requires all CCW lagoons to monitor groundwater for migrating pollutants. *Only three states* out of 15 require CCW landfills to be lined. It is not surprising, therefore, that EPA reported in 2000 that only 57 percent of CCW landfills and only 26% of CCW surface impoundments were lined and that only 65% of landfills and 38% of surface impoundments conducted groundwater monitoring.[21]

In addition, in 2005, a report prepared for EPA's Office of Solid Waste, entitled *Estimation of Costs for Regulating Fossil Fuel Combustion Ash Management at Large Electric Utilities Under Part 258*, included a survey on state disposal regulations that verified that states fail to prohibit the most dangerous CCW disposal practices. The report examined the top 25 coal-consuming states to determine how much CCW is prohibited from disposal below the natural water table. Since isolation of ash from water is critical to preventing toxic leachate, it is axiomatic that disposal of ash must occur *above* the water table. Yet the report found that only 16% of the total waste volume being regulated by these 25 states is prohibited from disposal in water when waste is disposed in surface impoundments. For landfills, the total waste volume that is prohibited from disposal in water is only 25%. Thus the great majority of total CCW produced in those states is allowed to be disposed into the water table, namely 84% of the total volume of CCW disposed in surface impoundments and 75% of the total volume disposed in landfills.[22]

In view of EPA's risk assessment that finds the cancer risk from ash ponds 900- times EPA's regulatory goals, the absence of basic monitoring, lining and isolation requirements at the nation's roughly 300 CCW surface impoundments is alarming. Failure to impose requirements at waste lagoons is particularly dangerous, because CCW disposed in surface impoundments is intentionally mixed with water to create a sludge. The presence of water facilitates the dissolution and migration of pollutants, particularly when the ash pond is unlined or lined with only soil or clay. As the above statistics reveal, lining and monitoring does occur at some CCW disposal units, but far too much is left to the discretion of state regulators and the whim of individual utilities.

A 2005 report published jointly by EPA and the U.S. Department of Energy (DOE), entitled "*Coal Combustion Waste Management at Landfills and Surface Impoundments, 1994-2004*, attempted to show that certain industry practices have improved since EPA's regulatory determination. The report was based primarily on data voluntarily submitted by the utility industry. The report surveyed 56 permitted landfills and surface impoundments built between 1994 and 2004. The report cited the presence of "liners" at all newly permitted surface impoundments and landfills and concluded "[t]he use of liners has become essentially ubiquitous." This conclusion, however, is grossly misleading, because the devil is in the details. While more liners appear to be installed on disposal units built in the last decade, the type of liners is insufficient to protect health and the environment.

In fact, the DOE/EPA Report reveals that only 39% of the units, at best, installed composite liners. According to EPA's 2007 draft *Human and Ecological Risk Assessment*, landfills and surface impoundments with clay liners do *not* provide adequate protection of health and the environment. EPA's *Risk Assessment* states:

> Risks from clay-lined units are lower than those from unlined units, but 90th percentile risks are still well above the risk criteria for arsenic and thallium for landfills and arsenic, boron and molybdenum for surface impoundments.[23]

The *Risk Assessment* further states that *composite liners* effectively reduce risks from all constituents to below the risk criteria for both landfills and surface impoundments. A composite liner is defined as a high-density polyethylene (HDPE) membrane combined with either geosynthetic or natural clays. Yet the DOE/EPA Report reveals that clay liners were used at 25% of the permitted units. Single liners, also deemed inadequate, were used at 18% of the surveyed units. Thus it is clear that the *majority* of new units do not have adequate liners. Unless the liner is of a sufficient quality to prevent the migration of contaminants, its use is largely irrelevant. The DOE/EPA Report's updated survey of state-permitted disposal units does not show that adequate protections are in place. Conversely, it reveals that the absence of a federal rule requiring composite liners has produced a whole new generation of waste units in at least a dozen states that pose serious threats to human health and the environment.

Furthermore, the 2005 DOE/EPA Report documents that nearly a third of the net disposable CCW generated in the U.S. are potentially *totally exempt* from solid waste permitting requirements.[24] The DOE/EPA Report explains this fact in great detail:

> [t]he six States that have solid waste permitting exemptions for certain on- site CCW landfills generated a total of approximately 17 million tons of net disposable CCWs in 2004, which is 20% of the total net disposable CCWs generated for all States. The one State that excludes CCW from all solid waste regulations, Alabama, generated a total of approximately 2.7 million tons of net disposable CCWs in 2004, which is about 3.3% of the total net disposable CCWs generated in all States. Ohio, which excludes "nontoxic" fly ash, bottom ash, and boiler slag from solid waste regulations, generated a total of 5.9 million tons of these wastes and 1.1 million tons of FGD wastes (about 7 million tons total) in 2004. Of these amounts, about 1.3 million tons of "nontoxic" fly ash, bottom ash, and boiler slag are

> beneficially used and about 1 million tons of FGD sludge are beneficially used. Hence, the net disposable CCWs that were potentially exempt from solid waste permitting requirements in Ohio in 2004 .... amount to about 4.6 million tons. .... Thus the amount of net disposable CCWs in Ohio that is potentially exempt from solid waste permitting requirements represents about 5.4% of the total net disposable CCWs generated for all States. **Overall, the portion of the net disposable CCWs that is potentially exempt from solid waste permitting requirements is approximately 24 million tons, which corresponds to 29% of the total net disposable CCWs generated in the United States during 2004.**[25]

The report also explains that this exempted CCW represents almost a third of the US coal-fired generating capacity:

> In terms of electric generating capacity, the six States that have solid waste permitting exemptions for certain on-site CCW landfills generated a total of approximately 66,000 MW, which is approximately 20% of the total coal-fired electric generating capacity in the United States in 2004. The one State the excluded CCWs from all solid waste regulations, Alabama, generated a total of approximately 12,000 MW in 2004, which is about 3.7% of the total. Ohio which excludes "nontoxic" fly ash, bottom ash and boiler slag from solid waste regulations, generated a total of about 24,000 MW in 2004. This represents about 7.2% of the total coal-fired electric generating capacity in the United States. **Overall, the portion of the coal-fired electric generating capacity in the States that potentially exempt CCW landfills from solid waste permitting requirements and that exclude certain CCWs from all solid waste regulation is approximately 102,000 MW, which corresponds to about 30% of the total coal-fired electric generating capacity in the United States in 2004.**[26]

(Emphasis added.) Thus the DOE/EPA Report demonstrates that a significant portion of the CCW generated in the U.S. is potentially not subject to *any* solid waste permitting. This is another wholly unacceptable gap in regulation of CCW that is likely to have significant negative impact on health and the environment.

6. Voluntary Industry Agreements are not a Solution

It is not viable to allow the utility industry to police itself. The proliferation of contaminated sites over the last 8 years demonstrates that industry is not voluntarily ensuring safe disposal. A voluntary agreement recently signed by some utilities and presented to EPA as a substitute for enforceable regulations is unacceptable.[27] Its shortcomings are too numerous to describe here in detail, but suffice it to say that the utilities are proposing substantially less protection for their toxic ash than is required by law for the garbage from their cafeterias.

The voluntary industry agreement is designed to allow the electric utility industry to continue avoiding the cost of safe disposal of its voluminous waste. The plan intentionally fails to require monitoring that would detect pollution escaping CCW surface impoundments and landfills or to require any specific response should pollution be detected. The plan fails to require the most basic of safeguards, composite liners, and it fails to prohibit the placement of CCW directly into groundwater, and nothing in the plan applies to disposal of CCW in mines. In view of continuing damage from coal ash, the hundreds of disposal units operated by industry today without safeguards, and the clear direction provided by Congress, the Clinton EPA and the National Academies of Science, it is untenable for any federal agency to entertain an unenforceable, voluntary proposal.

7. EPA Fails to Fulfill the Statutory Mandates of RCRA

The goal of RCRA is to ensure the safe disposal of solid and hazardous waste and to encourage the safe reuse of waste in order to protect human health and the environment and conserve the nation's natural resources.[28] By failing to make good on its promise to promulgate minimum federal standards, EPA has failed in both respects. The disposal of CCW without safeguards has resulted in the creation of "open dumps," as they are defined in 40 C.F.R. Part 257, which is specifically prohibited by the statute.[29] Furthermore, because disposal of CCW in unlined, unmonitored pits so frequently presents the threat of an imminent and substantial endangerment to health or the environment, these disposal units violate RCRA's core statutory mandate that disposal of solid waste avoid the potential for substantial damage, as set forth in section 7003 of RCRA.[30] Furthermore, Section 1008 of RCRA requires EPA to "develop and publish suggested guidelines" for solid waste management under subtitle D, as necessary to ensure protection of public health and the environment. Thus EPA has failed with regard to CCW, not only to abide by its own regulatory determination, but also to comply with the mandates of RCRA.

Further, by failing to impose disposal standards, EPA fails to encourage CCW reuse. When cheap dumping is no longer available, power plants will have far greater incentive to recycle their ash. Reuse of ash as a component of asphalt, concrete, and gypsum board are legitimate and safe reuses that should be encouraged. In addition, recycling ash in concrete can result in a large reduction of greenhouse gases. Approximately one ton of $CO2$ is released for every ton of Portland cement produced, but certain fly ashes can replace up to 50% by mass of Portland cement.[31] Further, since cement kilns are one of the largest emitters of mercury in the nation, the reduction of Portland cement production will reduce mercury emissions.

In Wisconsin, for example, adequate regulation of CCW has raised recycling rates significantly. Wisconsin CCW regulations are probably the most comprehensive in the nation. As a result, the recycling rate in Wisconsin for CCW is 85%, more than double the average recycling rate for all other CCW-producing states (36%).[32] It stands to reason that if the true cost of disposal were borne by electric utilities, there would be far greater incentive to find beneficial uses for the ash.

The Federal Solution

The solution is straightforward. EPA, or in the case of CCW disposal in mines, OSM, *in conjunction with EPA*, must provide minimum enforceable safeguards for the disposal of CCW in mines, landfills and waste lagoons. This is not a novel concept. These regulations can be similar to the regulations governing municipal solid waste landfills. For coal ash landfills, it is a simple matter to require the basics: placement above the water table, composite liners, groundwater monitoring, daily cover of the waste, cleanup standards if contamination is discovered, construction of a cap upon closure, financial assurance, and post-closure care. In fact, a coalition of environmental groups, including Earthjustice, submitted draft regulations to EPA almost 18 months ago. EPA never responded.

For disposal of coal ash in mines, the National Academies of Science established a clear framework for federal regulations in their 2006 report, recommending waste and site characterization, isolation from groundwater, effective monitoring, site specific management plans, adequate bonding, public participation in permitting, and site specific cleanup standards. Again, these basic safeguards are the familiar foundation of federal waste disposal law.

Recommendations

Many complicated environmental issues have been brought before this committee, but the instant question is not one of them. Clear solutions exist and have already been identified. Research and analysis conducted by EPA, the Science Advisory Board, and the National Academies of Science indicate a high and unacceptable risk from CCW when the waste is disposed without safeguards. The threat is not theoretical. Case after case of serious injury to health and the environment has resulted from unsafe disposal of CCW.

It is thus our hope that the Subcommittee will recommend that EPA and OSM take the following steps to protect our communities and environment from the risks posed by CCW.

1. A Timetable is Needed for Establishing Federal Regulations

For landfills and surface impoundments, EPA must immediately begin to formulate the basic minimum waste management requirements that will be required at all surface impoundments and landfills.

For standards applicable to mines, EPA should work closely with OSM. As necessary, RCRA authority must extend to waste disposal in mines, if it is found that SMCRA authority is not sufficient. Use of EPA's extensive expertise in waste management is essential to the development of effective and comprehensive waste disposal rules for mines, whether the regulations are promulgated under RCRA or SMCRA. EPA's decision to defer entirely to OSM and its consequent failure to work closely with OSM to ensure the quality of minefilling regulations is totally unacceptable.

In view of EPA's longstanding failure to abide by its 2000 commitment to promulgate regulations and the harm that is currently occurring because of EPA's failure to act, it is necessary to ensure that the agency is indeed moving forward to establish federal standards. Further action by this Subcommittee to conduct additional hearings and support legislation to set a deadline for federal action would be extremely helpful.

2. EPA and OSM Must Promulgate Federal Regulations, Not Guidance

We ask the Subcommittee to ensure that EPA and OSM establish regulations, not guidance, governing CCW disposal. Promulgation of federal *regulations* is absolutely essential, because many states cannot enact CCW disposal safeguards in the absence of federal standards. Some 23 states have "no more stringent" provisions in their statutes that prohibit the states from enacting stricter standards than are found in federal law. Thus for those states, without federal regulation, *there can be no regulation of CCW beyond what few safeguards there are now.*[33] Among states with "no more stringent provisions" are Colorado, Kentucky, Montana, New Mexico, Tennessee and Texas. While agency guidance is a useful tool to direct the implementation of enforceable regulations, it is not an acceptable substitute for a federal rulemaking.

3. EPA Should Phase-out Surface Impoundments (Waste Ponds) at Existing Coal-fired Plants and Prohibit the Construction of Surface Impoundments at New Plants

EPA should prohibit construction of surface impoundments at all new coal-fired plants and require a phasing-out of surface impoundments at existing plants. Electric utilities have a choice of producing dry or wet waste, and given the evidence of damage to human health and the environment from disposal of slurried (wet) ash in waste ponds, an essential and important step to improve waste management over the long term is to require utilities to move toward dry disposal of CCW. The dozens of cases of contamination from the leaching of arsenic and other pollutants from surface impoundments across the U.S. is testament to the danger of wet disposal. As described in this testimony, EPA's 2007 draft *Human and Ecological Risk Assessment of Coal Combustion Wastes* identifies exceedingly high risks of groundwater contamination from CCW surface impoundments and finds that the risk from surface impoundments is considerably higher than the risk from CCW landfills. Isolation of CCW from water is unquestionably the safest way to dispose of ash. A prohibition on new surface impoundments would greatly reduce the risk of new cases of poisoning and would ensure that waste management practices at the numerous new coal plants coming on line reflect our scientific knowledge. This prohibition would guarantee long-term protection because CCW waste units, particularly surface impoundments, are routinely used for several decades. Communities living near coal-fired power plants deserve protection from this wholly avoidable threat to their health and environment. For existing plants, EPA should establish reasonable date for termination of all wet-waste disposal. As an added benefit, disposing of dry ash in landfills preserves the ash for recycling at a later date.

4. EPA Should Prohibit Disposal of CCW in Sand and Gravel Pits

In view of the clear threat to public health posed by disposal of CCW in sand and gravel pits, we ask this Subcommittee to recommend an immediate prohibition. Since 2000, EPA has recommended that CCW disposal in sand and gravel pits be terminated because of the many damage cases resulting from this practice. Recently, CCW disposed in an unlined pit caused serious contamination of drinking water at the Gambrills site in Maryland. The threat to public health posed by the recent dumping (1999 through 2007) is unconscionable, considering EPA's long experience with cases of water contamination from this disposal practice. EPA has long acknowledged numerous proven damage cases caused by CCW disposal in sand and gravel pits, including sites that poisoned or threatened public drinking water supplies in Massachusetts, Virginia, and three sites in Wisconsin. A prohibition is necessary because this dangerous mode of disposal is still an acceptable practice in numerous states. In fact, Iowa currently has at least four ongoing disposal operations in unlined sand and gravel pits. Once again, EPA's scientific findings must be applied in a timely way to prevent future harm. In view of CCW's propensity to leach into aquifers from sand and gravel pits and the likely paths of migration to residential areas and public water supplies, it is necessary to act immediately to avoid further injury.

5.EPA Should Reject Voluntary Industry Proposals as a Substitute for Regulation

EPA must not consider a voluntary plan proposed by the utility industry as a substitute for regulations. If the utility industry is interested in moving forward with waste management improvements prior to EPA's adoption of regulations, that is commendable. Under no circumstances, however, should EPA consider such voluntary measures an acceptable substitute for national regulation.

CONCLUSION

In conclusion, Mr. Chairman, Representative Pearce and Members of the Subcommittee, Earthjustice asks the Subcommittee to ensure the promulgation of science-based, minimum federal standards, the hallmark of EPA's waste management program, to address the threat posed by coal combustion waste disposal. EPA and the National Research Council recognize, as does Congress, that mismanagement of CCW causes serious injury to public health and the environment. Maintenance of the status quo ensures that further damage will occur.

A great number of communities in the U.S. are concerned about this issue. OSM's *Advanced Notice of Proposed Rulemaking on the Placement of Coal Combustion Byproducts in Active and Abandoned Coal Mines* drew over 4,000 comments from citizens last June, and over 10,000 individuals responded to EPA's *Notice of Data Availability on Coal Combustion Wastes* in February 2008. Communities threatened by the disposal of coal ash are requesting that minimum standards be put in place as soon as possible. These communities, often poor and already fighting environmental threats from other sources, need to be protected from damage that is wholly preventable.

In its final *Regulatory Determination on Wastes from the Combustion of Fossil Fuels*, EPA determined that the cost to industry of compliance with tailored hazardous waste regulations would be "only a small percentage of industry revenues."[34] EPA estimated this cost to be "less than 0.4 percent of industry sales."[35] Today, EPA is considering regulating

CCW under solid waste authority, not under the far more costly subtitle C requirements of RCRA. Thus in 2005, EPA recalculated the cost to industry in its report, *Estimation of Costs for Regulating Fossil Fuel Combustion Ash Management at Large Electric Utilities Under Part 258*. EPA concluded that compliance with nonhazardous solid waste regulations would be less than half of the cost of compliance with hazardous waste rules.[36] Thus the cost of safe disposal is *not* burdensome to industry, although it has proved, at site after site, to be catastrophic to the public and the environment.

In sum, I greatly appreciate the Subcommittee's interest in the risk of harm posed by CCW and how this problem can be solved by our federal agencies. Thank you again, Mr. Chairman, for the opportunity to present to you and the Subcommittee information about this critical issue.

End Notes

[1] 65 Fed. Reg. 32214, May 22, 2000.

[2] U.S. EPA (1999). Report to Congress, Wastes from the combustion of fossil fuels. Volume 2 – Methods, findings, and recommendations. Office of Solid Waste and Emergency Response, Washington, DC. EPA 530-R-99-010. March 1999.

[3] U.S. Department of Energy (2004). Coal Combustion Waste Management at Landfill and Surface Impoundments 1994-2004. DOE/PI-004, ANL-EVS/06-4 at page 3.

[4] U.S. EPA (1999). Report to Congress, Wastes from the combustion of fossil fuels. Volume 2 – Methods, findings, and recommendations. Office of Solid Waste and Emergency Response, Washington, DC. EPA 530-R-99-010. March 1999.

[5] Annual Energy Outlook, 2007 with Projections to 2030 (Early Release)- Overview. Report No. DOE/EIA-0383/2007, December 2006.

[6] U.S. EPA (2006). Characterization of Mercury-Enriched Coal Combustion Residues from Electric Utilities Using Enhanced Sorbents for Mercury Control. EPA/600/R-06/008. (January).

[7] Ibid.

[8] U.S. EPA, Office of Research and Development. "Evaluating the Fate of Metals from Management of Coal Combustion Residues from Implementation of Multi-pollutant Controls at Coal-fired Electric Utilities," Presentation for 32nd Annual EPA-A&WMA Information Exchange. December 4, 2007.

[9] Adriano, D.C., Page, A.L., Elseewi, A.A., Chang, A.C., Straughan, I.R. (1980).Utilization and disposal of fly ash and other coal residues in terrestrial ecosystems. Journal of Environmental Quality, 9: 333.

[10] Carlson, C.L., Adriano, D.C. (1993). Environmental impacts of coal combustion residues. Journal of Environmental Quality, 22: 227-247.

[11] Cherry, D.S. (1999). A review of the adverse environmental impacts of coal combustion wastes. Prepared for the Hoosier Environmental Council, November 10, 1999.

[12] Taugher, Mike. Water Worries & Shumway Arroyo Was At Center of 1 980s Lawsuits, Albuquerque Journal, A1 & A10, October 24, 1999.

[13] Skorupa, Joseph, P. (1998). Selenium poisoning of fish and wildlife in nature: Lessons from twelve real- world examples, from Environmental Chemistry of Selenium. Marcel Dekker, Inc. New York.

[14] Cherry, D.S. (1999). A review of the adverse environmental impacts of coal combustion wastes. Prepared for the Hoosier Environmental Council, November 10, 1999

[15] Hopkins, W.A., C.L. Rowe, J.H. Roe, D.E.Scott, M.T. Mendon‡a and J.D. Congdon, (1999). Ecotoxicological impact of coal combustion byproducts on amphibians and reptiles. Savannah River Ecology Laboratory, presented at the Society for Environmental Toxicology and Chemistry, 20th Annual meeting, Philadelphia, PA, November 14-18. Abstract # PMP009.

[16] U.S. EPA (2002). Mine Placement of Coal Combustion Waste, State Program Elements, Final Draft, December 2002.

[17] Clean Air Task Force, Impacts of Waster Quality from Placement of Coal Combustion Waste in Pennsylvania Coal Mines, September 2007. Pages 26-38.www.catf.us

[18] See 40 C.F.R. Part 258.

[19] Ibid.

[20] 42 U.S.C. § 6947.

[21] US EPA. Regulatory Determination on Wastes from the Combustion of Fossil Fuels, 65 Fed. Reg. 32214 at 32216, May 22, 2000.
[22] DPRA Incorporated. Estimation of Costs for Regulating Fossil Fuel Combustion Ash Management at Large Electric Utilities under Part 258, prepared for U.S. EPA, Office of Solid Waste, November 30, 2005 at page 39.
[23] US EPA, Hunan and Ecological Risk Assessment at ES-7.
[24] U.S. Department of Energy (2004). Coal Combustion Waste Management at Landfill and Surface Impoundments 1994-2004. DOE/PI-004, ANL-EVS/06-4 at page 45.
[25] Ibid.
[26] Ibid. at 45-46.
[27] Utility Solid Waste Activities Group. Utility Industry Action Plan for the Management of Coal Combustion Products.
[28] 42 U.S.C. § 6902.
[29] 42 U.S.C. § 6945(a).
[30] 42 U.S.C. § 6973.
[31] http://www.us-concrete.com/news/features.asp (last checked June 1, 2008)
[32] U.S. Department of Energy (2004). Coal Combustion Waste Management at Landfill and Surface Impoundments 1994-2004. DOE/PI-004, ANL-EVS/06-4 at page 5.
[33] U.S. EPA (2002). Mine Placement of Coal Combustion Waste, State Program Elements, Final Draft, December 2002.
[34] 65 Fed. Reg. at 32230.
[35] Ibid.
[36] DPRA Incorporated. Estimation of Costs for Regulating Fossil Fuel Combustion Ash Management at Large Electric Utilities under Part 258, prepared for U.S. EPA, Office of Solid Waste, November 30, 2005.

In: Coal Combustion Waste: Management and Beneficial Uses
Editor: Daniel D. Lowell
ISBN: 978-1-61728-962-0

Chapter 16

WRITTEN TESTIMONY OF NORMAN K. HARVEY, PRESIDENT, GREATER GAMBRILLS IMPROVEMENT ASSOCIATION, BEFORE THE SUBCOMMITTEE ON ENERGY AND MINERAL RESOURCES, HEARING ON "HOW SHOULD THE FEDERAL GOVERNMENT ADDRESS THE HEALTH AND ENVIRONMENTAL RISKS OF COAL COMBUSTION WASTE?"

"How should the Federal Government address the Health and Environmental Risks of Coal Combustion Waste"?

I live in a very conservative multi-cultural neighborhood that was once predominantly African American. Being an African American and having been exposed to the many facets of public service, I was soon able to transfer skill sets and assistance to this small community that was besieged by large corporations and landfill operators. For decades these corporations had targeted them with disposal of chemical waste and toxic materials. Too often, and on a continuing basis, large organizations and businesses too eager to turn a large profit margin, target communities of disproportionate underrepresented minority groups (i.e. African Americans, Alaska Natives, American Indians, Mexican Americans and Hispanic groups) for chemical and toxic waste disposal.

Often focusing on certain areas of disparity in subject matter areas such as education, criminal and environmental justice, these corporations prey on these groups' socioeconomic status to unfairly take advantage of their communities, homes and lifestyles. The impact of these criminal predators is long felt months if not years later when health issues arise, and property and home values diminish. State and County officials who often work hand in hand to appease these perpetrators have either left office or attribute their decisions to the greater good of county revenue generated from taxes, permits and fees imposed. The Maryland Department of Environment (MDE), an agency charged to protect the environment and public health of its citizens, has consistently failed the very citizens that have been aggrieved in the Evergreen Road and Waugh Chapel communities.

In 1995 Constellation Energy in partnership with BBSS, Inc (i.e. Reliable Contractors) commenced depositing fly ash into a 63 acre pit known as the "Turner Pit" not more than one half mile from my community under the guise of reclamation but their real motive was profit. Fly ash is a byproduct of burned coal from power plants that capture it with air pollution control equipment. According to MDE, approximately 2 million tons of coal ash (i.e. fly ash and bottom ash, which is heavier than fly ash and is captured at the bottom of a combustion device) is currently generated each year in the State of Maryland. To date, more than 3.8 million tons of this highly toxic CCB have been dumped into the "Turner Pit" without adequate protective devices in place such as liners and leachate collection systems. In addition, the operators have neglected early warning signals from monitoring wells.

State and County officials were well aware of the probability of groundwater contamination due to earlier contamination of the Brandon Woods/Solley Road residential community seven years earlier. The Turner Pit project was approved by MDE just as the Brandon Woods project with the knowledge that severe ground water and aquifer contamination was an immediate threat of endangering the health of its citizens.

In June of 1999, MDE was made aware of groundwater and aquifer contamination above the 500 mg/l (milligrams per liter) permit limit for sulfates. MDE and the County Health Officer allowed continued operations; disregarding public health threats to the Evergreen and Waugh Chapel communities. Had a *site specific analysis* been conducted prior to commencing operations and the prior knowledge of the Brandon Woods/Solley Road disaster taken into consideration, this second incident of fly ash contamination would have been avoided by having the proper safeguards in place. A site analysis would have revealed that the acidity of the groundwater causes a greater acceptance to contamination without leachate systems and liners in place.

In 2000, MDE allowed special exceptions for extending the operation to an adjacent pit known as the "Waugh Chapel Pit"; disregarding specific findings and knowledge of leachate contamination at the first site including airborne contamination of fly ash particles. It has been estimated that the sites have generated more than $15 million dollars profit in taxes, permits and fees for Constellation Energy (a $19 billion dollar a year energy giant) and Reliable Contractors.

In 2004 a pump and treat system was installed at the "Turner Pit" to stem the tide of leachate plume down gradient from the site; however three residential wells already indicate high concentrations of calcium and potassium which are precursors to leachate migration, in addition to abnormal levels of aluminum being recorded. It is a fact that potentially cancer causing sulfates have been discharged into residential well water three times higher than EPA regulated safe standards.

In October 2006 test wells indicated 4,480 mgl/l at the Waugh Chapel site and operations were allowed to continue under existing MDE and County scrutiny. Some residents were forced to depend on garden hoses and pipes attached to fire hydrants for water, in addition to being furnished 12 ounce bottled water by Constellation Energy during the winter season.

In June 2007, according to a report by Maryland's Department of Natural Resources, 34 residential wells were contaminated with toxic elements including arsenic, lead and cadmium at levels as high as three times EPA's maximum standard for safe drinking water.

In September 2007 Constellation Energy voluntarily issued a "Consent Decree" to stop fly ash deposits after the site had been 90% filled, however the Waugh Chapel residents still rely on 12 ounce bottled water for their drinking needs.

Historically, across the nation and particularly in underrepresented areas, large waste producing corporations have sought old mining shafts, sand and gravel pits for fly ash disposal sites with little or no regard for public safety. The problem does not lie primarily with the waste product but the lack of proper safeguards and poorly regulated controls and ordinances.

In 1993 EPA under the Federal Resource Conservation and Recovery Act (RCRA) made a determination that it would no longer regulate coal combustion waste (i.e. coal ash/fly ash) as a hazardous waste. In doing so, EPA gave license to existing state waste managers who were not qualified nor equipped to adequately safeguard public safety or public drinking water.

Over the years, MDE has allowed loose interpretation of the EPA determination and non-enforcement of the same laws as necessary for industrial solid waste landfills. In addition, MDE fore-goes critical individual site review and environmental site analyses for permit applications. One study indicated that such reviews are critical for spotting potential hazardous conditions to communities and homeowners who are dependent on groundwater as a drinking water source or as in the case of the Crofton Area Township which relies on three aquifers for public drinking water as well.

It is and has been determined that MDE should have required a site liner and leachate collection systems prior to the Turner and Waugh Chapel operations, but again the Maryland watchdog agency in place failed to protect the very citizens who depended on them for their public safety and health.

Currently, MDE is in the process of re-writing proposals and regulations for stricter disposal of fly ash but without the participation of local citizen groups and environmental justice organizations most affected and that would benefit from such partnership. Environment Maryland along with Crofton 1st are organizations that would prove most beneficial to MDE's newly proposed Coal Combustion Byproduct regulations by working with the county and state to (1) allow community and public involvement in the rulemaking process and (2) ensure that new legal requirements covering the use of fly ash in landfills and abandoned mines are adequately protective for underrepresented communities that have long been the targets for chemical and toxic waste disposal.

The term, "beneficial use" of coal ash must be redefined from roadway fill, highway embankments, soil conditioner usage and with greater measures to ensure that it is mixed with or used as a bonding agent to prohibit environmental/public exposure.

Federal, State and County officials must safeguard the general public and any close lying communities from fly ash particles in ambient air. Open fields of fly ash particles generate clouds of dust often coating nearby residential homes and cars. Operating permits should include plans for monitoring coal dust and stringent enforcement.

Also, any existing CCB facilities should not be grandfathered or allowed to expand under old existing permits and/or granted special exceptions as currently and previously been the norm.

Once protective systems are in place (i.e. liners and leachate systems) it should be certified and verified by federal, state and/or county officials to ascertain that it meets all necessary requirements. Officials on site must guarantee correct liner thickness and proper placement prior to any fly ash deposits.

Most importantly, it is of my opinion that statutory mandates should be enacted instead of regulations now currently being developed or proposed. As seen in the past, regulations can often be administratively changed, but statutory mandates that are voted into law are not susceptible to quick change by administrative/county officials as is currently the case.

In closing, I would like to see a special delegate for future rulemaking processes in the A.A. County assembly and elected by the citizens with a defined role to raise environmental issues that have so grossly been ignored by State and County officials. Notwithstanding, I would also organize a citizen watchdog steering committee entitled PECCL (the People's Environmental Coalition for Cleaner Living) that would work with that special delegate and serve to ensure that these unfortunate events would indeed be a thing forever of the past for citizens in Anne Arundel County.

Norman K. Harvey

CHAPTER SOURCES

The following chapters have been previously published:

Chapter 1 – This is an edited, excerpted and augmented edition of a United States Congressional Research Service publication, Report Order Code R40544, dated January 12, 2010.

Chapter 2 – This is an edited, excerpted and augmented edition of a United States Environmental Protection Agency publication, dated February 12, 2008.

Chapter 3 – These remarks were delivered as Statement of Professor Avner Vengosh, before the United States House of Representatives Subcommittee on Water Resources and Environment, dated March 31, 2009.

Chapter 4 – These remarks were delivered as Statement of Tom Kilgore, before the United States House of Representatives Subcommittee on Water Resources and Environment, dated March 31, 2009.

Chapter 5 – These remarks were delivered as Statement of Stan Meiberg, before the United States House of Representatives Subcommittee on Water Resources and Environment, dated March 31, 2009.

Chapter 6 – These remarks were delivered as Statement of Barry Breen, before the United States House of Representatives Subcommittee on Water Resources and Environment, dated April 30, 2009.

Chapter 7 – These remarks were delivered as Statement of Eric Schaeffer and Lisa Evans, before the United States House of Representatives Subcommittee on Water Resources and Environment, dated April 30, 2009.

Chapter 8 – These remarks were delivered as Statement of David C. Goss, before the United States House of Representatives Subcommittee on Water Resources and Environment, dated April 30, 2009.

Chapter 9 – These remarks were delivered as Statement of Jim Costa, before the United States House of Representatives Subcommittee onEnergy and Mineral Resources, dated June 10, 2008.

Chapter 10 – These remarks were delivered as Statement of Mark Squillace, before the United States House of Representatives Subcommittee on Energy and Mineral Resources, dated June 10, 2008.

Chapter 11 – These remarks were delivered as Statement of Shari T. Wilson, before the United States House of Representatives Subcommittee on Energy and Mineral Resources, dated June 10, 2008.

Chapter 12 – These remarks were delivered as Statement of David Goss, before the United States House of Representatives Subcommittee on Energy and Mineral Resources, dated June 10, 2008.

Chapter 13 – These remarks were delivered as Statement of Charles H. Norris, before the United States House of Representatives Subcommittee on Energy and Mineral Resources, dated June 10, 2008.

Chapter 14 – These remarks were delivered as Statement of Mary A. Fox, PhD, MPH, before the United States House of Representatives Subcommittee on Energy and Mineral Resources, dated June 10, 2008.

Chapter 15 – These remarks were delivered as Statement of Lisa Evans, before the United States House of Representatives Subcommittee on Energy and Mineral Resources, dated June 10, 2008.

Chapter 16 - These remarks were delivered as Statement of Norman K. Harvey, before the United States House of Representatives Subcommittee on Energy and Mineral Resources, dated June 10, 2008.

INDEX

S

T

U

V

W

Z